Copyright © 2015 by Jonathan English
All rights reserved. This book or any portion thereof
may not be reproduced or used in any manner whatsoever
without the express written permission of the publisher
except for the use of brief quotations in a book review.
Printed in the United States of America

All individual texts included are in the public domain, except the new translations provided by the editor.

First Printing, 2015

ISBN 978-1-329-49956-0

Table of Contents

Reading 1 – Plato: *Meno* and *Timaeus* — 1

Reading 2 – Aristotle: *Physics* — 11

Reading 3 – Archimedes: *On the Equilibrium of Planes of The Centres of Gravity of Planes* — 24

Reading 4 – Galileo: *Two New Sciences* — 26

Reading 5 – Descartes: *Le Monde* — 40

Reading 6 – Newton: *Principia, Definitions* — 48

Reading 7 – Newton: *Principia, Axioms* — 58

Reading 8 – Maxwell: *Theory of Heat* — 66

Reading 9 – Huygens: *Treatise on Light* — 71

Reading 10 – Newton: *New Theory About Light and Colour* — 78

Reading 11 – Young: Lecture XXXIX. *On the Nature of Light and Colors* — 86

Reading 12 – Taylor: *De motu Nervi tensi* — 93

Reading 13 – Gilbert: *De Magnete* — 95

Reading 14 – Du Fay: *A Letter from Mons. Du Fay concerning Electricity.* — 102

Reading 15 – Franklin: *From Benjamin Franklin to Peter Collinson* — 105

Reading 16 – Faraday: *Experimental Researches in Electricity* — 107

Reading 17 – Einstein: *On the Electrodynamics of moving bodies* — 118

Reading 1
Plato: Meno and Timaeus

Meno (79e-86d)

Translation by Benjamin Jowett

SOCRATES: But then, my friend, do not suppose that we can explain to any one the nature of virtue as a whole through some unexplained portion of virtue, or anything at all in that fashion; we should only have to ask over again the old question, What is virtue? Am I not right?

MENO: I believe that you are.

5   SOCRATES: Then begin again, and answer me, What, according to you and your friend Gorgias, is the definition of virtue?

MENO: O Socrates, I used to be told, before I knew you, that you were always doubting yourself and making others doubt; and now you are casting your spells over me, and I am simply getting bewitched and enchanted, and am at my wits' end. And if I may venture to make a jest upon you,
10 you seem to me both in your appearance and in your power over others to be very like the flat torpedo fish, who torpifies those who come near him and touch him, as you have now torpified me, I think. For my soul and my tongue are really torpid, and I do not know how to answer you; and though I have been delivered of an infinite variety of speeches about virtue before now, and to many persons—and very good ones they were, as I thought—at this moment I cannot even say
15 what virtue is. And I think that you are very wise in not voyaging and going away from home, for if you did in other places as you do in Athens, you would be cast into prison as a magician.

SOCRATES: You are a rogue, Meno, and had all but caught me.

MENO: What do you mean, Socrates?

SOCRATES: I can tell why you made a simile about me.

20 MENO: Why?

SOCRATES: In order that I might make another simile about you. For I know that all pretty young gentlemen like to have pretty similes made about them—as well they may—but I shall not return the compliment. As to my being a torpedo, if the torpedo is torpid as well as the cause of torpidity in others, then indeed I am a torpedo, but not otherwise; for I perplex others, not because
25 I am clear, but because I am utterly perplexed myself. And now I know not what virtue is, and you seem to be in the same case, although you did once perhaps know before you touched me. However, I have no objection to join with you in the enquiry.

MENO: And how will you enquire, Socrates, into that which you do not know? What will you put forth as the subject of enquiry? And if you find what you want, how will you ever know that this is
30 the thing which you did not know?

SOCRATES: I know, Meno, what you mean; but just see what a tiresome dispute you are introducing. You argue that a man cannot enquire either about that which he knows, or about that which he does not know; for if he knows, he has no need to enquire; and if not, he cannot; for he does not know the very subject about which he is to enquire (Compare Aristot. Post. Anal.).

35 MENO: Well, Socrates, and is not the argument sound?

SOCRATES: I think not.

MENO: Why not?

SOCRATES: I will tell you why: I have heard from certain wise men and women who spoke of things divine that—

MENO: What did they say?

SOCRATES: They spoke of a glorious truth, as I conceive.

MENO: What was it? and who were they?

SOCRATES: Some of them were priests and priestesses, who had studied how they might be able to give a reason of their profession: there have been poets also, who spoke of these things by inspiration, like Pindar, and many others who were inspired. And they say—mark, now, and see whether their words are true—they say that the soul of man is immortal, and at one time has an end, which is termed dying, and at another time is born again, but is never destroyed. And the moral is, that a man ought to live always in perfect holiness. 'For in the ninth year Persephone sends the souls of those from whom she has received the penalty of ancient crime back again from beneath into the light of the sun above, and these are they who become noble kings and mighty men and great in wisdom and are called saintly heroes in after ages.' The soul, then, as being immortal, and having been born again many times, and having seen all things that exist, whether in this world or in the world below, has knowledge of them all; and it is no wonder that she should be able to call to remembrance all that she ever knew about virtue, and about everything; for as all nature is akin, and the soul has learned all things; there is no difficulty in her eliciting or as men say learning, out of a single recollection all the rest, if a man is strenuous and does not faint; for all enquiry and all learning is but recollection. And therefore we ought not to listen to this sophistical argument about the impossibility of enquiry: for it will make us idle; and is sweet only to the sluggard; but the other saying will make us active and inquisitive. In that confiding, I will gladly enquire with you into the nature of virtue.

MENO: Yes, Socrates; but what do you mean by saying that we do not learn, and that what we call learning is only a process of recollection? Can you teach me how this is?

SOCRATES: I told you, Meno, just now that you were a rogue, and now you ask whether I can teach you, when I am saying that there is no teaching, but only recollection; and thus you imagine that you will involve me in a contradiction.

MENO: Indeed, Socrates, I protest that I had no such intention. I only asked the question from habit; but if you can prove to me that what you say is true, I wish that you would.

SOCRATES: It will be no easy matter, but I will try to please you to the utmost of my power. Suppose that you call one of your numerous attendants, that I may demonstrate on him.

MENO: Certainly. Come hither, boy.

SOCRATES: He is Greek, and speaks Greek, does he not?

MENO: Yes, indeed; he was born in the house.

SOCRATES: Attend now to the questions which I ask him, and observe whether he learns of me or only remembers.

MENO: I will.

SOCRATES: Tell me, boy, do you know that a figure like this is a square?

BOY: I do.

SOCRATES: And you know that a square figure has these four lines equal?

BOY: Certainly.

80 SOCRATES: And these lines which I have drawn through the middle of the square are also equal?

BOY: Yes.

SOCRATES: A square may be of any size?

BOY: Certainly.

85 SOCRATES: And if one side of the figure be of two feet, and the other side be of two feet, how much will the whole be? Let me explain: if in one direction the space was of two feet, and in the other direction of one foot, the whole would be of two feet taken once?

BOY: Yes.

SOCRATES: But since this side is also of two feet, there are twice two feet?

90 BOY: There are.

SOCRATES: Then the square is of twice two feet?

BOY: Yes.

SOCRATES: And how many are twice two feet? Count and tell me.

BOY: Four, Socrates.

95 SOCRATES: And might there not be another square twice as large as this, and having like this the lines equal?

BOY: Yes.

SOCRATES: And of how many feet will that be?

BOY: Of eight feet.

100 SOCRATES: And now try and tell me the length of the line which forms the side of that double square: this is two feet—what will that be?

BOY: Clearly, Socrates, it will be double.

SOCRATES: Do you observe, Meno, that I am not teaching the boy anything, but only asking him questions; and now he fancies that he knows how long a line is necessary in order to produce a
105 figure of eight square feet; does he not?

MENO: Yes.

SOCRATES: And does he really know?

MENO: Certainly not.

SOCRATES: He only guesses that because the square is double, the line is double.

110 MENO: True.

SOCRATES: Observe him while he recalls the steps in regular order. (To the Boy:) Tell me, boy, do you assert that a double space comes from a double line? Remember that I am not speaking of an oblong, but of a figure equal every way, and twice the size of this—that is to say of eight feet; and I want to know whether you still say that a double square comes from double line?

115 BOY: Yes.

SOCRATES: But does not this line become doubled if we add another such line here?

BOY: Certainly.

SOCRATES: And four such lines will make a space containing eight feet?

BOY: Yes.

SOCRATES: Let us describe such a figure: Would you not say that this is the figure of eight feet?

BOY: Yes.

SOCRATES: And are there not these four divisions in the figure, each of which is equal to the figure of four feet?

BOY: True.

SOCRATES: And is not that four times four?

BOY: Certainly.

SOCRATES: And four times is not double?

BOY: No, indeed.

SOCRATES: But how much?

BOY: Four times as much.

SOCRATES: Therefore the double line, boy, has given a space, not twice, but four times as much.

BOY: True.

SOCRATES: Four times four are sixteen—are they not?

BOY: Yes.

SOCRATES: What line would give you a space of eight feet, as this gives one of sixteen feet;—do you see?

BOY: Yes.

SOCRATES: And the space of four feet is made from this half line?

BOY: Yes.

SOCRATES: Good; and is not a space of eight feet twice the size of this, and half the size of the other?

BOY: Certainly.

SOCRATES: Such a space, then, will be made out of a line greater than this one, and less than that one?

BOY: Yes; I think so.

SOCRATES: Very good; I like to hear you say what you think. And now tell me, is not this a line of two feet and that of four?

BOY: Yes.

SOCRATES: Then the line which forms the side of eight feet ought to be more than this line of two feet, and less than the other of four feet?

BOY: It ought.

SOCRATES: Try and see if you can tell me how much it will be.

BOY: Three feet.

SOCRATES: Then if we add a half to this line of two, that will be the line of three. Here are two and there is one; and on the other side, here are two also and there is one: and that makes the figure of which you speak?

BOY: Yes.

SOCRATES: But if there are three feet this way and three feet that way, the whole space will be three times three feet?

BOY: That is evident.

SOCRATES: And how much are three times three feet?

BOY: Nine.

SOCRATES: And how much is the double of four?

165 BOY: Eight.

SOCRATES: Then the figure of eight is not made out of a line of three?

BOY: No.

SOCRATES: But from what line?—tell me exactly; and if you would rather not reckon, try and show me the line.

170 BOY: Indeed, Socrates, I do not know.

SOCRATES: Do you see, Meno, what advances he has made in his power of recollection? He did not know at first, and he does not know now, what is the side of a figure of eight feet: but then he thought that he knew, and answered confidently as if he knew, and had no difficulty; now he has a difficulty, and neither knows nor fancies that he knows.

175 MENO: True.

SOCRATES: Is he not better off in knowing his ignorance?

MENO: I think that he is.

SOCRATES: If we have made him doubt, and given him the 'torpedo's shock,' have we done him any harm?

180 MENO: I think not.

SOCRATES: We have certainly, as would seem, assisted him in some degree to the discovery of the truth; and now he will wish to remedy his ignorance, but then he would have been ready to tell all the world again and again that the double space should have a double side.

MENO: True.

185 SOCRATES: But do you suppose that he would ever have enquired into or learned what he fancied that he knew, though he was really ignorant of it, until he had fallen into perplexity under the idea that he did not know, and had desired to know?

MENO: I think not, Socrates.

SOCRATES: Then he was the better for the torpedo's touch?

190 MENO: I think so.

SOCRATES: Mark now the farther development. I shall only ask him, and not teach him, and he shall share the enquiry with me: and do you watch and see if you find me telling or explaining anything to him, instead of eliciting his opinion. Tell me, boy, is not this a square of four feet which I have drawn?

195 BOY: Yes.

SOCRATES: And now I add another square equal to the former one?

BOY: Yes.

SOCRATES: And a third, which is equal to either of them?

BOY: Yes.

200 SOCRATES: Suppose that we fill up the vacant corner?

BOY: Very good.

SOCRATES: Here, then, there are four equal spaces?

BOY: Yes.

SOCRATES: And how many times larger is this space than this other?

BOY: Four times.

SOCRATES: But it ought to have been twice only, as you will remember.

BOY: True.

SOCRATES: And does not this line, reaching from corner to corner, bisect each of these spaces?

BOY: Yes.

SOCRATES: And are there not here four equal lines which contain this space?

BOY: There are.

SOCRATES: Look and see how much this space is.

BOY: I do not understand.

SOCRATES: Has not each interior line cut off half of the four spaces?

BOY: Yes.

SOCRATES: And how many spaces are there in this section?

BOY: Four.

SOCRATES: And how many in this?

BOY: Two.

SOCRATES: And four is how many times two?

BOY: Twice.

SOCRATES: And this space is of how many feet?

BOY: Of eight feet.

SOCRATES: And from what line do you get this figure?

BOY: From this.

SOCRATES: That is, from the line which extends from corner to corner of the figure of four feet?

BOY: Yes.

SOCRATES: And that is the line which the learned call the diagonal. And if this is the proper name, then you, Meno's slave, are prepared to affirm that the double space is the square of the diagonal?

BOY: Certainly, Socrates.

SOCRATES: What do you say of him, Meno? Were not all these answers given out of his own head?

MENO: Yes, they were all his own.

SOCRATES: And yet, as we were just now saying, he did not know?

MENO: True.

SOCRATES: But still he had in him those notions of his—had he not?

MENO: Yes.

SOCRATES: Then he who does not know may still have true notions of that which he does not know?

MENO: He has.

SOCRATES: And at present these notions have just been stirred up in him, as in a dream; but if he were frequently asked the same questions, in different forms, he would know as well as any one at last?

MENO: I dare say.

SOCRATES: Without any one teaching him he will recover his knowledge for himself, if he is only asked questions?

MENO: Yes.

SOCRATES: And this spontaneous recovery of knowledge in him is recollection?

MENO: True.

SOCRATES: And this knowledge which he now has must he not either have acquired or always possessed?

MENO: Yes.

SOCRATES: But if he always possessed this knowledge he would always have known; or if he has acquired the knowledge he could not have acquired it in this life, unless he has been taught geometry; for he may be made to do the same with all geometry and every other branch of knowledge. Now, has any one ever taught him all this? You must know about him, if, as you say, he was born and bred in your house.

MENO: And I am certain that no one ever did teach him.

SOCRATES: And yet he has the knowledge?

MENO: The fact, Socrates, is undeniable.

SOCRATES: But if he did not acquire the knowledge in this life, then he must have had and learned it at some other time?

MENO: Clearly he must.

SOCRATES: Which must have been the time when he was not a man?

MENO: Yes.

SOCRATES: And if there have been always true thoughts in him, both at the time when he was and was not a man, which only need to be awakened into knowledge by putting questions to him, his soul must have always possessed this knowledge, for he always either was or was not a man?

MENO: Obviously.

SOCRATES: And if the truth of all things always existed in the soul, then the soul is immortal. Wherefore be of good cheer, and try to recollect what you do not know, or rather what you do not remember.

MENO: I feel, somehow, that I like what you are saying.

SOCRATES: And I, Meno, like what I am saying. Some things I have said of which I am not altogether confident. But that we shall be better and braver and less helpless if we think that we ought to enquire, than we should have been if we indulged in the idle fancy that there was no knowing and no use in seeking to know what we do not know;—that is a theme upon which I am ready to fight, in word and deed, to the utmost of my power.

MENO: There again, Socrates, your words seem to me excellent.

SOCRATES: Then, as we are agreed that a man should enquire about that which he does not know, shall you and I make an effort to enquire together into the nature of virtue?

MENO: By all means, Socrates. And yet I would much rather return to my original question, Whether in seeking to acquire virtue we should regard it as a thing to be taught, or as a gift of nature, or as coming to men in some other way?

Timaeus (27b-30c, 57e-58c)

Translation by R. D. Archer-Hind

SOKRATES. Ample and splendid indeed, it seems, will be the banquet of discourse which I am to receive in my turn. So it would seem to be your business to speak next, Timaeus, after you have duly invoked the gods.

TIMAEUS. Yes indeed, Sokrates, that is what all do who possess the slightest share of judgment; at the outset of every work, great or small, they always call upon a god: and seeing that we are going to enter on a discussion of the universe, how far it is created or perchance uncreate, unless we are altogether beside ourselves, we must needs invoke the gods and goddesses and pray above all that our discourse may be pleasing in their sight, next that it may be consistent with itself. Let it suffice then thus to have called upon the gods; but we must call upon ourselves likewise to conduct the discourse in such a way that you will most readily comprehend me, and I shall most fully carry out my intentions in expounding the subject that is before us.

First then in my judgment this distinction must be made. What is that which is eternally and has no becoming, and again what is that which comes to be but is never? The one is comprehensible by thought with the aid of reason, ever changeless; the other opinable by opinion with the aid of reasonless sensation, becoming and perishing, never truly existent. Now all that comes to be must needs be brought into being by some cause: for it is impossible for anything without a cause to attain to birth. Of whatsoever thing then the Artificer, looking ever to the changeless and using that as his model, works out the design and function, all that is so accomplished must needs be fair: but if he look to that which has come to be, using the created as his model, the work is not fair. Now as to the whole heaven or order of the universe for whatsoever name is most acceptable to it, be it so named by us we must first ask concerning it the question which lies at the outset of every inquiry, whether did it exist eternally, having no beginning of generation, or has it come into being, starting from some beginning? It has come into being: for it can be seen and felt and has body; and all such things are sensible, and sensible things, apprehensible by opinion with sensation, belong, as we saw, to becoming and creation. We say that what has come to be must be brought into being by some cause. Now the maker and father of this All it were a hard task to find, and having found him, it were impossible to declare him to all men. However we must again inquire concerning him, after which of the models did the framer of it fashion the universe, after the changeless and abiding, or after that which has come into being? If now this universe is fair and its Artificer good, it is plain that he looked to the eternal; but if nay it may not even be uttered without impiety, then it was to that which has come into being. Now it is manifest to everyone that he looked to the eternal: for the universe is fairest of all things that have come to be, and he is the most excellent of causes. And having come on this wise into being it has been created in the image of that which is comprehensible by reason and wisdom and changes never. Granting this, it must needs be that this universe is a likeness of something. Now it is all-important to make our beginning according to nature: and this affirmation must be laid down with regard to a likeness and its model, that the words must be akin to the subjects of which they are the interpreters : therefore of that which is abiding and sure and discoverable by the aid of reason the words too must be abiding and unchanging, and so far as it lies in words to be incontrovertible and immovable, they must in no wise fall short of this ; but those which deal with that which is made in the image of the former and which is a likeness must be likely and duly corresponding with their subject : as being is to becoming, so is truth to belief. If then, Sokrates, after so many men have said divers things concerning the gods and the generation of the universe, we should not prove able to render an account everywhere and in all respects consistent and accurate, let no one be surprised; but if we can produce one as probable as any other, we must be content, remembering that I who speak and you my judges are but men: so that on these subjects we should be satisfied with the probable story and seek nothing further.

SOKRATES. Quite right, Timaeus; we must accept it exactly as you say. Your prelude is exceedingly welcome to us, so please proceed with the strain itself.

TIMAEUS. Let us declare then for what cause nature and this All was framed by him that framed it. He was good, and in none that is good can there arise jealousy of aught at any time. So being far aloof from this, he desired that all things should be as like unto himself as possible. This is that most sovereign cause of nature and the universe which we shall most surely be right in accepting from men of understanding. For God desiring that all things should be good, and that, so far as this might be, there should be naught evil, having received all that is visible not in a state of rest, but moving without harmony or measure, brought it from its disorder into order, thinking that this was in all ways better than the other. Now it neither has been nor is permitted to the most perfect to do aught but what is most fair. Therefore he took thought and perceived that of all things which are by nature visible, no work that is without reason will ever be fairer than that which has reason, setting whole against whole, and that without soul reason cannot dwell in anything. Because then he argued thus, in forming the universe he created reason in soul and soul in body, that he might be the maker of a work that was by nature most fair and perfect. In this way then we ought to affirm according to the probable account that this universe is a living creature in very truth possessing soul and reason by the providence of God.

…

Now concerning rest and motion, how they arise and under what conditions, we must come to an agreement, else many difficulties will stand in the way of our argument that is to follow. This has been already in part set forth, but we have yet to add that in uniformity no movement will ever exist. For that what is to be moved should exist without that which is to move it, or what is to move without that which is to be moved, is difficult or rather impossible : but without these there can be no motion, and for these to be uniform is not possible. So then let us always assign rest to uniformity and motion to its opposite. Now the opposite of uniformity is caused by in-equality; and of inequality we have discussed the origin. But how it comes to pass that all bodies are not sorted off into their several kinds and cease from passing through one another and changing their place, this we have not explained. Let us put it again in this way. The revolution of the whole, when it had embraced the four kinds, being circular, with a natural tendency to return upon itself, compresses everything and suffers no vacant space to be left. Therefore fire penetrates most of all through all things, and in the second degree air, since it is second in fineness, and the rest in proportion. For the substances which are formed of the largest parts have the most void left in their structure, and those made of the smallest have the least. Now the constriction of this contracting force thrusts the small particles into the interspaces between the larger: so that when small are set side by side with great, and the lesser particles divide the greater, while the greater compress the smaller, all things keep rushing backwards and forwards to their own region; since in changing its bulk each changes its proper position in space. Thus owing to these causes a perpetual disturbance of uniformity is always kept up and so preserves the perpetual motion of matter now and henceforth without cessation.

Reading 2
Aristotle: Physics

Physics Book III Part 1

Translated By R. P. Hardie and R. K. Gaye

Nature has been defined as a 'principle of motion and change', and it is the subject of our inquiry. We must therefore see that we understand the meaning of 'motion'; for if it were unknown, the meaning of 'nature' too would be unknown.

When we have determined the nature of motion, our next task will be to attack in the same way the terms which are involved in it. Now motion is supposed to belong to the class of things which are continuous; and the infinite presents itself first in the continuous-that is how it comes about that 'infinite' is often used in definitions of the continuous ('what is infinitely divisible is continuous'). Besides these, place, void, and time are thought to be necessary conditions of motion.

Clearly, then, for these reasons and also because the attributes mentioned are common to, and coextensive with, all the objects of our science, we must first take each of them in hand and discuss it. For the investigation of special attributes comes after that of the common attributes.

To begin then, as we said, with motion.
We may start by distinguishing (1) what exists in a state of fulfilment only, (2) what exists as potential, (3) what exists as potential and also in fulfilment-one being a 'this', another 'so much', a third 'such', and similarly in each of the other modes of the predication of being.

Further, the word 'relative' is used with reference to (1) excess and defect, (2) agent and patient and generally what can move and what can be moved. For 'what can cause movement' is relative to 'what can be moved', and vice versa.

Again, there is no such thing as motion over and above the things. It is always with respect to substance or to quantity or to quality or to place that what changes changes. But it is impossible, as we assert, to find anything common to these which is neither 'this' nor quantum nor quale nor any of the other predicates. Hence neither will motion and change have reference to something over and above the things mentioned, for there is nothing over and above them.

Now each of these belongs to all its subjects in either of two ways: namely (1) substance-the one is positive form, the other privation; (2) in quality, white and black; (3) in quantity, complete and incomplete; (4) in respect of locomotion, upwards and downwards or light and heavy. Hence there are as many types of motion or change as there are meanings of the word 'is'.

We have now before us the distinctions in the various classes of being between what is full real and what is potential.

405   Def. The fulfilment of what exists potentially, in so far as it exists potentially, is motion-namely, of what is alterable qua alterable, alteration: of what can be increased and its opposite what can be decreased (there is no common name), increase and decrease: of what can come to be and can pass away, coming to he and passing away: of what can be carried along, locomotion.

Examples will elucidate this definition of motion. When the buildable, in so far as it is just that, is
410 fully real, it is being built, and this is building. Similarly, learning, doctoring, rolling, leaping, ripening, ageing.

The same thing, if it is of a certain kind, can be both potential and fully real, not indeed at the same time or not in the same respect, but e.g. potentially hot and actually cold. Hence at once such things will act and be acted on by one another in many ways: each of them will be capable at the
415 same time of causing alteration and of being altered. Hence, too, what effects motion as a physical agent can be moved: when a thing of this kind causes motion, it is itself also moved. This, indeed, has led some people to suppose that every mover is moved. But this question depends on another set of arguments, and the truth will be made clear later is possible for a thing to cause motion, though it is itself incapable of being moved.

420   It is the fulfilment of what is potential when it is already fully real and operates not as itself but as movable, that is motion. What I mean by 'as' is this: Bronze is potentially a statue. But it is not the fulfilment of bronze as bronze which is motion. For 'to be bronze' and 'to be a certain potentiality' are not the same.

If they were identical without qualification, i.e. in definition, the fulfilment of bronze as bronze
425 would have been motion. But they are not the same, as has been said. (This is obvious in contraries. 'To be capable of health' and 'to be capable of illness' are not the same, for if they were there would be no difference between being ill and being well. Yet the subject both of health and of sickness-whether it is humour or blood-is one and the same.)

We can distinguish, then, between the two-just as, to give another example, 'colour' and visible'
430 are different-and clearly it is the fulfilment of what is potential as potential that is motion. So this, precisely, is motion.

Further it is evident that motion is an attribute of a thing just when it is fully real in this way, and neither before nor after. For each thing of this kind is capable of being at one time actual, at another not. Take for instance the buildable as buildable. The actuality of the buildable as buildable is the
435 process of building. For the actuality of the buildable must be either this or the house. But when there is a house, the buildable is no longer buildable. On the other hand, it is the buildable which is being built. The process then of being built must be the kind of actuality required. But building is a kind of motion, and the same account will apply to the other kinds also.

Book III Part 2

The soundness of this definition is evident both when we consider the accounts of motion that
440 the others have given, and also from the difficulty of defining it otherwise.

One could not easily put motion and change in another genus-this is plain if we consider where some people put it; they identify motion with or 'inequality' or 'not being'; but such things are not necessarily moved, whether they are 'different' or 'unequal' or 'non-existent'; Nor is change either to or from these rather than to or from their opposites.

The reason why they put motion into these genera is that it is thought to be something indefinite, and the principles in the second column are indefinite because they are privative: none of them is either 'this' or 'such' or comes under any of the other modes of predication. The reason in turn why motion is thought to be indefinite is that it cannot be classed simply as a potentiality or as an actuality-a thing that is merely capable of having a certain size is not undergoing change, nor yet a thing that is actually of a certain size, and motion is thought to be a sort of actuality, but incomplete, the reason for this view being that the potential whose actuality it is is incomplete. This is why it is hard to grasp what motion is. It is necessary to class it with privation or with potentiality or with sheer actuality, yet none of these seems possible. There remains then the suggested mode of definition, namely that it is a sort of actuality, or actuality of the kind described, hard to grasp, but not incapable of existing.

The mover too is moved, as has been said-every mover, that is, which is capable of motion, and whose immobility is rest-when a thing is subject to motion its immobility is rest. For to act on the movable as such is just to move it. But this it does by contact, so that at the same time it is also acted on. Hence we can define motion as the fulfilment of the movable qua movable, the cause of the attribute being contact with what can move so that the mover is also acted on. The mover or agent will always be the vehicle of a form, either a 'this' or 'such', which, when it acts, will be the source and cause of the change, e.g. the full-formed man begets man from what is potentially man.

Book III Part 3

The solution of the difficulty that is raised about the motion-whether it is in the movable-is plain. It is the fulfilment of this potentiality, and by the action of that which has the power of causing motion; and the actuality of that which has the power of causing motion is not other than the actuality of the movable, for it must be the fulfilment of both. A thing is capable of causing motion because it can do this, it is a mover because it actually does it. But it is on the movable that it is capable of acting. Hence there is a single actuality of both alike, just as one to two and two to one are the same interval, and the steep ascent and the steep descent are one-for these are one and the same, although they can be described in different ways. So it is with the mover and the moved.

This view has a dialectical difficulty. Perhaps it is necessary that the actuality of the agent and that of the patient should not be the same. The one is 'agency' and the other 'patiency'; and the outcome and completion of the one is an 'action', that of the other a 'passion'. Since then they are both motions, we may ask: in what are they, if they are different? Either (a) both are in what is acted on and moved, or (b) the agency is in the agent and the patiency in the patient. (If we ought to call the latter also 'agency', the word would be used in two senses.)

Now, in alternative (b), the motion will be in the mover, for the same statement will hold of 'mover' and 'moved'. Hence either every mover will be moved, or, though having motion, it will not be moved.

If on the other hand (a) both are in what is moved and acted on-both the agency and the patiency (e.g. both teaching and learning, though they are two, in the learner), then, first, the actuality of each will not be present in each, and, a second absurdity, a thing will have two motions at the same time. How will there be two alterations of quality in one subject towards one definite quality? The thing is impossible: the actualization will be one.

But (someone will say) it is contrary to reason to suppose that there should be one identical actualization of two things which are different in kind. Yet there will be, if teaching and learning are the same, and agency and patiency. To teach will be the same as to learn, and to act the same as to be acted on-the teacher will necessarily be learning everything that he teaches, and the agent will be acted on. One may reply:

(1) It is not absurd that the actualization of one thing should be in another. Teaching is the activity of a person who can teach, yet the operation is performed on some patient-it is not cut adrift from a subject, but is of A on B.

(2) There is nothing to prevent two things having one and the same actualization, provided the actualizations are not described in the same way, but are related as what can act to what is acting.

(3) Nor is it necessary that the teacher should learn, even if to act and to be acted on are one and the same, provided they are not the same in definition (as 'raiment' and 'dress'), but are the same merely in the sense in which the road from Thebes to Athens and the road from Athens to Thebes are the same, as has been explained above. For it is not things which are in a way the same that have all their attributes the same, but only such as have the same definition. But indeed it by no means follows from the fact that teaching is the same as learning, that to learn is the same as to teach, any more than it follows from the fact that there is one distance between two things which are at a distance from each other, that the two vectors AB and Ba, are one and the same. To generalize, teaching is not the same as learning, or agency as patiency, in the full sense, though they belong to the same subject, the motion; for the 'actualization of X in Y' and the 'actualization of Y through the action of X' differ in definition.

What then Motion is, has been stated both generally and particularly. It is not difficult to see how each of its types will be defined-alteration is the fulfillment of the alterable qua alterable (or, more scientifically, the fulfilment of what can act and what can be acted on, as such)-generally and again in each particular case, building, healing, &c. A similar definition will apply to each of the other kinds of motion.

Book IV Part 1

The physicist must have a knowledge of Place, too, as well as of the infinite-namely, whether there is such a thing or not, and the manner of its existence and what it is-both because all suppose that things which exist are somewhere (the non-existent is nowhere--where is the goat-stag or the sphinx?), and because 'motion' in its most general and primary sense is change of place, which we call 'locomotion'.

The question, what is place? presents many difficulties. An examination of all the relevant facts seems to lead to divergent conclusions. Moreover, we have inherited nothing from previous thinkers, whether in the way of a statement of difficulties or of a solution.

The existence of place is held to be obvious from the fact of mutual replacement. Where water now is, there in turn, when the water has gone out as from a vessel, air is present. When therefore another body occupies this same place, the place is thought to be different from all the bodies which come to be in it and replace one another. What now contains air formerly contained water, so that clearly the place or space into which and out of which they passed was something different from both.

Further, the typical locomotions of the elementary natural bodies-namely, fire, earth, and the like- show not only that place is something, but also that it exerts a certain influence. Each is carried to its own place, if it is not hindered, the one up, the other down. Now these are regions or kinds of place-up and down and the rest of the six directions. Nor do such distinctions (up and down and right and left, &c.) hold only in relation to us. To us they are not always the same but change with the direction in which we are turned: that is why the same thing may be both right and left, up and down, before and behind. But in nature each is distinct, taken apart by itself. It is not every chance direction which is 'up', but where fire and what is light are carried; similarly, too, 'down' is not any chance direction but where what has weight and what is made of earth are carried-the implication being that these places do not differ merely in relative position, but also as possessing distinct potencies. This is made plain also by the objects studied by mathematics. Though they have no real place, they nevertheless, in respect of their position relatively to us, have a right and left as attributes ascribed to them only in consequence of their relative position, not having by nature these various characteristics. Again, the theory that the void exists involves the existence of place: for one would define void as place bereft of body.

These considerations then would lead us to suppose that place is something distinct from bodies, and that every sensible body is in place. Hesiod too might be held to have given a correct account of it when he made chaos first. At least he says: '*First of all things came chaos to being, then broad-breasted earth,*' implying that things need to have space first, because he thought, with most people, that everything is somewhere and in place. If this is its nature, the potency of place must be a marvelous thing, and take precedence of all other things. For that without which nothing else can exist, while it can exist without the others, must needs be first; for place does not pass out of existence when the things in it are annihilated.

True, but even if we suppose its existence settled, the question of its nature presents difficulty- whether it is some sort of 'bulk' of body or some entity other than that, for we must first determine its genus.

(1) Now it has three dimensions, length, breadth, depth, the dimensions by which all body also is bounded. But the place cannot be body; for if it were there would be two bodies in the same place.

(2) Further, if body has a place and space, clearly so too have surface and the other limits of body; for the same statement will apply to them: where the bounding planes of the water were, there in turn will be those of the air. But when we come to a point we cannot make a distinction between it and its place. Hence if the place of a point is not different from the point, no more will that of any of the others be different, and place will not be something different from each of them.

(3) What in the world then are we to suppose place to be? If it has the sort of nature described, it cannot be an element or composed of elements, whether these be corporeal or incorporeal: for

560 while it has size, it has not body. But the elements of sensible bodies are bodies, while nothing that has size results from a combination of intelligible elements.

(4) Also we may ask: of what in things is space the cause? None of the four modes of causation can be ascribed to it. It is neither in the sense of the matter of existents (for nothing is composed of it), nor as the form and definition of things, nor as end, nor does it move existents.

565 (5) Further, too, if it is itself an existent, where will it be? Zeno's difficulty demands an explanation: for if everything that exists has a place, place too will have a place, and so on ad infinitum.

(6) Again, just as every body is in place, so, too, every place has a body in it. What then shall we say about growing things? It follows from these premises that their place must grow with them, if their place is neither less nor greater than they are.

570 By asking these questions, then, we must raise the whole problem about place-not only as to what it is, but even whether there is such a thin Part 4 g. It will now be plain from these considerations what place is. There are just four things of which place must be one-the shape, or the matter, or some sort of extension between the bounding surfaces of the containing body, or this boundary itself if it contains no extension over and above the bulk of the body which comes to be in it.

575 Three of these it obviously cannot be:
(1) The shape is supposed to be place because it surrounds, for the extremities of what contains and of what is contained are coincident. Both the shape and the place, it is true, are boundaries. But not of the same thing: the form is the boundary of the thing, the place is the boundary of the body which contains it.

580 (2) The extension between the extremities is thought to be something, because what is contained and separate may often be changed while the container remains the same (as water may be poured from a vessel)-the assumption being that the extension is something over and above the body displaced. But there is no such extension. One of the bodies which change places and are naturally capable of being in contact with the container falls in whichever it may chance to be. If there were
585 an extension which were such as to exist independently and be permanent, there would be an infinity of places in the same thing. For when the water and the air change places, all the portions of the two together will play the same part in the whole which was previously played by all the water in the vessel; at the same time the place too will be undergoing change; so that there will be another place which is the place of the place, and many places will be coincident. There is not a
590 different place of the part, in which it is moved, when the whole vessel changes its place: it is always the same: for it is in the (proximate) place where they are that the air and the water (or the parts of the water) succeed each other, not in that place in which they come to be, which is part of the place which is the place of the whole world.

595 (3) The matter, too, might seem to be place, at least if we consider it in what is at rest and is thus separate but in continuity. For just as in change of quality there is something which was formerly black and is now white, or formerly soft and now hard-this is just why we say that the matter exists-so place, because it presents a similar phenomenon, is thought to exist-only in the one case we say so because what was air is now water, in the other because where air formerly was there a is now

water. But the matter, as we said before, is neither separable from the thing nor contains it, whereas place has both characteristics.

Well, then, if place is none of the three-neither the form nor the matter nor an extension which is always there, different from, and over and above, the extension of the thing which is displaced-place necessarily is the one of the four which is left, namely, the boundary of the containing body at which it is in contact with the contained body. (By the contained body is meant what can be moved by way of locomotion.)

Place is thought to be something important and hard to grasp, both because the matter and the shape present themselves along with it, and because the displacement of the body that is moved takes place in a stationary container, for it seems possible that there should be an interval which is other than the bodies which are moved. The air, too, which is thought to be incorporeal, contributes something to the belief: it is not only the boundaries of the vessel which seem to be place, but also what is between them, regarded as empty. Just, in fact, as the vessel is transportable place, so place is a non-portable vessel. So when what is within a thing which is moved, is moved and changes its place, as a boat on a river, what contains plays the part of a vessel rather than that of place. Place on the other hand is rather what is motionless: so it is rather the whole river that is place, because as a whole it is motionless.

Hence we conclude that the innermost motionless boundary of what contains is place.

This explains why the middle of the heaven and the surface which faces us of the rotating system are held to be 'up' and 'down' in the strict and fullest sense for all men: for the one is always at rest, while the inner side of the rotating body remains always coincident with itself. Hence since the light is what is naturally carried up, and the heavy what is carried down, the boundary which contains in the direction of the middle of the universe, and the middle itself, are down, and that which contains in the direction of the outermost part of the universe, and the outermost part itself, are up.

For this reason, too, place is thought to be a kind of surface, and as it were a vessel, i.e. a container of the thing.

Further, place is coincident with the thing, for boundaries are coincident with the bounded.

Book IV Part 10

Next for discussion after the subjects mentioned is Time. The best plan will be to begin by working out the difficulties connected with it, making use of the current arguments. First, does it belong to the class of things that exist or to that of things that do not exist? Then secondly, what is its nature? To start, then: the following considerations would make one suspect that it either does not exist at all or barely, and in an obscure way. One part of it has been and is not, while the other is going to be and is not yet. Yet time-both infinite time and any time you like to take-is made up of these. One would naturally suppose that what is made up of things which do not exist could have no share in reality.

Further, if a divisible thing is to exist, it is necessary that, when it exists, all or some of its parts must exist. But of time some parts have been, while others have to be, and no part of it is though it is divisible. For what is 'now' is not a part: a part is a measure of the whole, which must be made up of parts. Time, on the other hand, is not held to be made up of 'nows'.

Again, the 'now' which seems to bound the past and the future-does it always remain one and the same or is it always other and other? It is hard to say.

(1) If it is always different and different, and if none of the parts in time which are other and other are simultaneous (unless the one contains and the other is contained, as the shorter time is by the longer), and if the 'now' which is not, but formerly was, must have ceased-to-be at some time, the 'nows' too cannot be simultaneous with one another, but the prior 'now' must always have ceased-to-be. But the prior 'now' cannot have ceased-to-be in itself (since it then existed); yet it cannot have ceased-to-be in another 'now'. For we may lay it down that one 'now' cannot be next to another, any more than point to point. If then it did not cease-to-be in the next 'now' but in another, it would exist simultaneously with the innumerable 'nows' between the two-which is impossible.

Yes, but (2) neither is it possible for the 'now' to remain always the same. No determinate divisible thing has a single termination, whether it is continuously extended in one or in more than one dimension: but the 'now' is a termination, and it is possible to cut off a determinate time. Further, if coincidence in time (i.e. being neither prior nor posterior) means to be 'in one and the same "now"', then, if both what is before and what is after are in this same 'now', things which happened ten thousand years ago would be simultaneous with what has happened to-day, and nothing would be before or after anything else.

This may serve as a statement of the difficulties about the attributes of time.

As to what time is or what is its nature, the traditional accounts give us as little light as the preliminary problems which we have worked through.

Some assert that it is (1) the movement of the whole, others that it is (2) the sphere itself.

(1) Yet part, too, of the revolution is a time, but it certainly is not a revolution: for what is taken is part of a revolution, not a revolution. Besides, if there were more heavens than one, the movement of any of them equally would be time, so that there would be many times at the same time.

(2) Those who said that time is the sphere of the whole thought so, no doubt, on the ground that all things are in time and all things are in the sphere of the whole. The view is too naive for it to be worthwhile to consider the impossibilities implied in it.

But as time is most usually supposed to be (3) motion and a kind of change, we must consider this view.

Now (a) the change or movement of each thing is only in the thing which changes or where the thing itself which moves or changes may chance to be. But time is present equally everywhere and with all things.

Again, (b) change is always faster or slower, whereas time is not: for 'fast' and 'slow' are defined by time-'fast' is what moves much in a short time, 'slow' what moves little in a long time; but time is not defined by time, by being either a certain amount or a certain kind of it.

Clearly then it is not movement. (We need not distinguish at present between 'movement' and 'change'.)

Book IV Part 11

But neither does time exist without change; for when the state of our own minds does not change at all, or we have not noticed its changing, we do not realize that time has elapsed, any more than those who are fabled to sleep among the heroes in Sardinia do when they are awakened; for they connect the earlier 'now' with the later and make them one, cutting out the interval because of their failure to notice it. So, just as, if the 'now' were not different but one and the same, there would not have been time, so too when its difference escapes our notice the interval does not seem to be time. If, then, the non-realization of the existence of time happens to us when we do not distinguish any change, but the soul seems to stay in one indivisible state, and when we perceive and distinguish we say time has elapsed, evidently time is not independent of movement and change. It is evident, then, that time is neither movement nor independent of movement.

We must take this as our starting-point and try to discover-since we wish to know what time is-what exactly it has to do with movement.

Now we perceive movement and time together: for even when it is dark and we are not being affected through the body, if any movement takes place in the mind we at once suppose that some time also has elapsed; and not only that but also, when some time is thought to have passed, some movement also along with it seems to have taken place. Hence time is either movement or something that belongs to movement. Since then it is not movement, it must be the other.

But what is moved is moved from something to something, and all magnitude is continuous. Therefore the movement goes with the magnitude. Because the magnitude is continuous, the movement too must be continuous, and if the movement, then the time; for the time that has passed is always thought to be in proportion to the movement.

The distinction of 'before' and 'after' holds primarily, then, in place; and there in virtue of relative position. Since then 'before' and 'after' hold in magnitude, they must hold also in movement, these corresponding to those. But also in time the distinction of 'before' and 'after' must hold, for time and movement always correspond with each other. The 'before' and 'after' in motion is identical in substratum with motion yet differs from it in definition, and is not identical with motion.

But we apprehend time only when we have marked motion, marking it by 'before' and 'after'; and it is only when we have perceived 'before' and 'after' in motion that we say that time has elapsed. Now we mark them by judging that A and B are different, and that some third thing is intermediate to them. When we think of the extremes as different from the middle and the mind pronounces that the 'nows' are two, one before and one after, it is then that we say that there is time, and this that we say is time. For what is bounded by the 'now' is thought to be time-we may assume this.

When, therefore, we perceive the 'now' one, and neither as before and after in a motion nor as an identity but in relation to a 'before' and an 'after', no time is thought to have elapsed, because there has been no motion either. On the other hand, when we do perceive a 'before' and an 'after', then we say that there is time. For time is just this-number of motion in respect of 'before' and 'after'.

Hence time is not movement, but only movement in so far as it admits of enumeration. A proof of this: we discriminate the more or the less by number, but more or less movement by time. Time then is a kind of number. (Number, we must note, is used in two senses-both of what is counted or the countable and also of that with which we count. Time obviously is what is counted, not that with which we count: there are different kinds of thing.) Just as motion is a perpetual succession, so also is time. But every simultaneous time is self-identical; for the 'now' as a subject is an identity, but it accepts different attributes. The 'now' measures time, in so far as time involves the 'before and after'.

The 'now' in one sense is the same, in another it is not the same. In so far as it is in succession, it is different (which is just what its being was supposed to mean), but its substratum is an identity: for motion, as was said, goes with magnitude, and time, as we maintain, with motion. Similarly, then, there corresponds to the point the body which is carried along, and by which we are aware of the motion and of the 'before and after' involved in it. This is an identical substratum (whether a point or a stone or something else of the kind), but it has different attributes as the sophists assume that Coriscus' being in the Lyceum is a different thing from Coriscus' being in the market-place. And the body which is carried along is different, in so far as it is at one time here and at another there. But the 'now' corresponds to the body that is carried along, as time corresponds to the motion. For it is by means of the body that is carried along that we become aware of the 'before and after' the motion, and if we regard these as countable we get the 'now'. Hence in these also the 'now' as substratum remains the same (for it is what is before and after in movement), but what is predicated of it is different; for it is in so far as the 'before and after' is numerable that we get the 'now'. This is what is most knowable: for, similarly, motion is known because of that which is moved, locomotion because of that which is carried. What is carried is a real thing, the movement is not. Thus what is called 'now' in one sense is always the same; in another it is not the same: for this is true also of what is carried.

Clearly, too, if there were no time, there would be no 'now', and vice versa. Just as the moving body and its locomotion involve each other mutually, so too do the number of the moving body and the number of its locomotion. For the number of the locomotion is time, while the 'now' corresponds to the moving body, and is like the unit of number.

Time, then, also is both made continuous by the 'now' and divided at it. For here too there is a correspondence with the locomotion and the moving body. For the motion or locomotion is made one by the thing which is moved, because it is one-not because it is one in its own nature (for there might be pauses in the movement of such a thing)-but because it is one in definition: for this determines the movement as 'before' and 'after'. Here, too there is a correspondence with the point; for the point also both connects and terminates the length-it is the beginning of one and the end of another. But when you take it in this way, using the one point as two, a pause is necessary, if the same point is to be the beginning and the end. The 'now' on the other hand, since the body carried is moving, is always different.

Hence time is not number in the sense in which there is 'number' of the same point because it is beginning and end, but rather as the extremities of a line form a number, and not as the parts of the line do so, both for the reason given (for we can use the middle point as two, so that on that analogy time might stand still), and further because obviously the 'now' is no part of time nor the section any

755 part of the movement, any more than the points are parts of the line-for it is two lines that are parts of one line.

In so far then as the 'now' is a boundary, it is not time, but an attribute of it; in so far as it numbers, it is number; for boundaries belong only to that which they bound, but number (e.g. ten) is the number of these horses, and belongs also elsewhere.

760 It is clear, then, that time is 'number of movement in respect of the before and after', and is continuous since it is an attribute of what is continuous.

Book VIII Part 1

It remains to consider the following question. Was there ever a becoming of motion before which it had no being, and is it perishing again so as to leave nothing in motion? Or are we to say that it never had any becoming and is not perishing, but always was and always will be? Is it in fact
765 an immortal never-failing property of things that are, a sort of life as it were to all naturally constituted things?

Now the existence of motion is asserted by all who have anything to say about nature, because they all concern themselves with the construction of the world and study the question of becoming and perishing, which processes could not come about without the existence of motion. But those
770 who say that there is an infinite number of worlds, some of which are in process of becoming while others are in process of perishing, assert that there is always motion (for these processes of becoming and perishing of the worlds necessarily involve motion), whereas those who hold that there is only one world, whether everlasting or not, make corresponding assumptions in regard to motion. If then it is possible that at any time nothing should be in motion, this must come about in
775 one of two ways: either in the manner described by Anaxagoras, who says that all things were together and at rest for an infinite period of time, and that then Mind introduced motion and separated them; or in the manner described by Empedocles, according to whom the universe is alternately in motion and at rest-in motion, when Love is making the one out of many, or Strife is making many out of one, and at rest in the intermediate periods of time-his account being as
780 follows: '*Since One hath learned to spring from Manifold, And One disjoined makes manifold arise, Thus they Become, nor stable is their life: But since their motion must alternate be, Thus have they ever Rest upon their round*': for we must suppose that he means by this that they alternate from the one motion to the other. We must consider, then, how this matter stands, for the discovery of the truth about it is of importance, not only for the study of nature, but also for the investigation of the First Principle.

785 Let us take our start from what we have already laid down in our course on Physics. Motion, we say, is the fulfilment of the movable in so far as it is movable. Each kind of motion, therefore, necessarily involves the presence of the things that are capable of that motion. In fact, even apart from the definition of motion, everyone would admit that in each kind of motion it is that which is capable of that motion that is in motion: thus it is that which is capable of alteration that is altered,
790 and that which is capable of local change that is in locomotion: and so there must be something capable of being burned before there can be a process of being burned, and something capable of burning before there can be a process of burning. Moreover, these things also must either have a beginning before which they had no being, or they must be eternal. Now if there was a becoming of every movable thing, it follows that before the motion in question another change or motion must

have taken place in which that which was capable of being moved or of causing motion had its becoming. To suppose, on the other hand, that these things were in being throughout all previous time without there being any motion appears unreasonable on a moment's thought, and still more unreasonable, we shall find, on further consideration. For if we are to say that, while there are on the one hand things that are movable, and on the other hand things that are motive, there is a time when there is a first movent and a first moved, and another time when there is no such thing but only something that is at rest, then this thing that is at rest must previously have been in process of change: for there must have been some cause of its rest, rest being the privation of motion. Therefore, before this first change there will be a previous change. For some things cause motion in only one way, while others can produce either of two contrary motions: thus fire causes heating but not cooling, whereas it would seem that knowledge may be directed to two contrary ends while remaining one and the same. Even in the former class, however, there seems to be something similar, for a cold thing in a sense causes heating by turning away and retiring, just as one possessed of knowledge voluntarily makes an error when he uses his knowledge in the reverse way. But at any rate all things that are capable respectively of affecting and being affected, or of causing motion and being moved, are capable of it not under all conditions, but only when they are in a particular condition and approach one another: so it is on the approach of one thing to another that the one causes motion and the other is moved, and when they are present under such conditions as rendered the one motive and the other movable. So if the motion was not always in process, it is clear that they must have been in a condition not such as to render them capable respectively of being moved and of causing motion, and one or other of them must have been in process of change: for in what is relative this is a necessary consequence: e.g. if one thing is double another when before it was not so, one or other of them, if not both, must have been in process of change. It follows then, that there will be a process of change previous to the first.

(Further, how can there be any 'before' and 'after' without the existence of time? Or how can there be any time without the existence of motion? If, then, time is the number of motion or itself a kind of motion, it follows that, if there is always time, motion must also be eternal. But so far as time is concerned we see that all with one exception are in agreement in saying that it is uncreated: in fact, it is just this that enables Democritus to show that all things cannot have had a becoming: for time, he says, is uncreated. Plato alone asserts the creation of time, saying that it had a becoming together with the universe, the universe according to him having had a becoming. Now since time cannot exist and is unthinkable apart from the moment, and the moment a kind of middle-point, uniting as it does in itself both a beginning and an end, a beginning of future time and an end of past time, it follows that there must always be time: for the extremity of the last period of time that we take must be found in some moment, since time contains no point of contact for us except the moment. Therefore, since the moment is both a beginning and an end, there must always be time on both sides of it. But if this is true of time, it is evident that it must also be true of motion, time being a kind of affection of motion.)

The same reasoning will also serve to show the imperishability of motion: just as a becoming of motion would involve, as we saw, the existence of a process of change previous to the first, in the same way a perishing of motion would involve the existence of a process of change subsequent to the last: for when a thing ceases to be moved, it does not therefore at the same time cease to be movable-e.g. the cessation of the process of being burned does not involve the cessation of the capacity of being burned, since a thing may be capable of being burned without being in process of being burned-nor, when a thing ceases to be movent, does it therefore at the same time cease to a be motive. Again, the destructive agent will have to be destroyed, after what it destroys has been destroyed, and then that which has the capacity of destroying it will have to be destroyed afterwards, (so that there will be a process of change subsequent to the last,) for being destroyed also is a kind of change. If, then, view which we are criticizing involves these impossible consequences, it is clear that motion is eternal and cannot have existed at one time and not at another: in fact such a view can hardly be described as anything else than fantastic.

And much the same may be said of the view that such is the ordinance of nature and that this must be regarded as a principle, as would seem to be the view of Empedocles when he says that the constitution of the world is of necessity such that Love and Strife alternately predominate and cause motion, while in the intermediate period of time there is a state of rest. Probably also those who like Anaxagoras, assert a single principle (of motion) would hold this view. But that which is produced or directed by nature can never be anything disorderly: for nature is everywhere the cause of order. Moreover, there is no ratio in the relation of the infinite to the infinite, whereas order always means ratio. But if we say that there is first a state of rest for an infinite time, and then motion is started at some moment, and that the fact that it is this rather than a previous moment is of no importance, and involves no order, then we can no longer say that it is nature's work: for if anything is of a certain character naturally, it either is so invariably and is not sometimes of this and sometimes of another character (e.g. fire, which travels upwards naturally, does not sometimes do so and sometimes not) or there is a ratio in the variation. It would be better, therefore, to say with Empedocles and anyone else who may have maintained such a theory as his that the universe is alternately at rest and in motion: for in a system of this kind we have at once a certain order. But even here the holder of the theory ought not only to assert the fact: he ought to explain the cause of it: i.e. he should not make any mere assumption or lay down any gratuitous axiom, but should employ either inductive or demonstrative reasoning. The Love and Strife postulated by Empedocles are not in themselves causes of the fact in question, nor is it of the essence of either that it should be so, the essential function of the former being to unite, of the latter to separate. If he is to go on to explain this alternate predominance, he should adduce cases where such a state of things exists, as he points to the fact that among mankind we have something that unites men, namely Love, while on the other hand enemies avoid one another: thus from the observed fact that this occurs in certain cases comes the assumption that it occurs also in the universe. Then, again, some argument is needed to explain why the predominance of each of the two forces lasts for an equal period of time. But it is a wrong assumption to suppose universally that we have an adequate first principle in virtue of the fact that something always is so or always happens so. Thus Democritus reduces the causes that explain nature to the fact that things happened in the past in the same way as they happen now: but he does not think fit to seek for a first principle to explain this 'always': so, while his theory is right in so far as it is applied to certain individual cases, he is wrong in making it of universal application. Thus, a triangle always has its angles equal to two right angles, but there is nevertheless an ulterior cause of the eternity of this truth, whereas first principles are eternal and have no ulterior cause. Let this conclude what we have to say in support of our contention that there never was a time when there was not motion, and never will be a time when there will not be motion.

Reading 3

Archimedes: On the Equilibrium of Planes of The Centres of Gravity of Planes

Book 1

Translation by Sir Thomas Little Heath

"I POSTULATE the following:

1. Equal weights at equal distances are in equilibrium, and equal weights at unequal distances are not in equilibrium but incline towards the weight which is at the greater distance.

2. If, when weights at certain distances are in equilibrium, something be added to one of the weights, they are not in equilibrium but incline towards that weight to which the addition was made.

3. Similarly, if anything be taken away from one of the weights, they are not in equilibrium but incline towards the weight from which nothing was taken.

4. When equal and similar plane figures coincide if applied to one another, their centres of gravity similarly coincide.

5. In figures which are unequal but similar the centres of gravity will be similarly situated. By points similarly situated in relation to similar figures I mean points such that, if straight lines be drawn from them to the equal angles, they make equal angles with the corresponding sides.

6. If magnitudes at certain distances be in equilibrium, (other) magnitudes equal to them will also be in equilibrium at the same distances.

7. In any figure whose perimeter is concave in (one and) the same direction the centre of gravity must be within the figure."

Proposition 1.

Weights which balance at equal distances are equal.

For, if they are unequal, take away from the greater the difference between the two. The remainders will then not balance [Post. 3] ; which is absurd.

Therefore the weights cannot be unequal.

Proposition 2.

Unequal weights at equal distances will not balance but will incline towards the greater weight.

For take away from the greater the difference between the two. The equal remainders will therefore balance [Post. 1]. Hence, if we add the difference again, the weights will not balance but incline towards the greater [Post. 2].

Proposition 3.

Unequal weights will balance at unequal distances, the greater weight being at the lesser distance.

Let A, B be two unequal weights (of which A is the greater) balancing about C at distances AC, BC respectively. Then shall AC he less than BC. For, if not, take away from A the weight (A — B.) The remainders will then incline towards B [Post. 3]. But this is impossible, for (1) if AC = CB, the equal remainders will balance, or (2) AC> CB, they will incline towards A at the greater distance [Post. 1].

Hence AG<CB.

Conversely, if the weights balance, and AC < CB, then A>B.

Proposition 4.

If two equal weights have not the same centre of gravity, the centre of gravity of both taken together is at the middle point of the line joining their centres of gravity.

[Proved from Prop. 3 by *reductio ad absurdum*. Archimedes assumes that the centre of gravity of both together is on the straight line joining the centres of gravity of each, saying that this had been proved before. The allusion is no doubt to the lost treatise *On levers*.]

Proposition 14.

It follows at once from the last proposition that the centre of gravity of any triangle is at the intersection of the lines drawn from any two angles to the middle points of the opposite sides respectively.

Reading 4
Galileo: Two New Sciences

First Day: Interlocutors: Salviati, Sagredo and Simplicio

Translation by Henry Crew and Alfonso de Salvio

SALV.

The constant activity which you Venetians display in your famous arsenal suggests to the studious mind a large field for investigation, especially that part of the work which involves mechanics; for in this department all types of instruments and machines are constantly being constructed by many artisans, among whom there must be some who, partly by inherited experience and partly by their own observations, have become highly expert and clever in explanation.

SAGR.

You are quite right. Indeed, I myself, being curious by nature, frequently visit this place for the mere pleasure of observing the work of those who, on account of their superiority over other artisans, we call "first rank men." Conference with them has often helped me in the investigation of certain effects including not only those which are striking, but also those which are recondite and almost incredible. At times also I have been put to confusion and driven to despair of ever explaining something for which I could not account, but which my senses told me to be true. And notwithstanding the fact that what the old man told us a little while ago is proverbial and commonly accepted, yet it seemed to me altogether false, like many another saying which is current among the ignorant; for I think they introduce these expressions in order to give the appearance of knowing something about matters which they do not understand.

SALV.

You refer, perhaps, to that last remark of his when we asked the reason why they employed stocks, scaffolding and bracing of larger dimensions for launching a big vessel than they do for a small one; and he answered that they did this in order to avoid the danger of the ship parting under its own heavy weight [*vasta mole*], a danger to which small boats are not subject?

SAGR.

Yes, that is what I mean; and I refer especially to his last assertion which I have always regarded as a false, though current, opinion; namely, that in speaking of these and other similar machines one cannot argue from the small to the large, because many devices which succeed on a small scale do not work on a large scale. Now, since mechanics has its foundation in geometry, where mere size cuts no figure, I do not see that the properties of circles, triangles, cylinders, cones and other solid figures will change with their size. If, therefore, a large machine be constructed in such a way that its parts bear to one another the same ratio as in a smaller one, and if the smaller is sufficiently strong for the purpose for which it was designed, I do not see why the larger also should not be able to withstand any severe and destructive tests to which it may be subjected.

SALV.

The common opinion is here absolutely wrong. Indeed, it is so far wrong that precisely the opposite is true, namely, that many machines can be constructed even more perfectly on a large scale than on a small; thus, for instance, a clock which indicates and strikes the hour can be made more accurate on a large scale than on a small. There are some intelligent people who maintain this same opinion, but on more reasonable grounds, when they cut loose from geometry and argue that the better performance of the large machine is owing to the imperfections and variations of the material. Here I trust you will not charge me with arrogance if I say that imperfections in the material, even those which are great enough to invalidate the clearest mathematical proof, are not sufficient to explain the deviations observed between machines in the concrete and in the abstract. Yet I shall say it and will affirm that, even if the imperfections did not exist and matter were absolutely perfect, unalterable and free from all accidental variations, still the mere fact that it is matter makes the larger machine, built of the same material and in the same proportion as the smaller, correspond with exactness to the smaller in every respect except that it will not be so strong or so resistant against violent treatment; the larger the machine, the greater its weakness. Since I assume matter to be unchangeable and always the same, it is clear that we are no less able to treat this constant and invariable property in a rigid manner than if it belonged to simple and pure mathematics. Therefore, Sagredo, you would do well to change the opinion which you, and perhaps also many other students of mechanics, have entertained concerning the ability of machines and structures to resist external disturbances, thinking that when they are built of the same material and maintain the same ratio between parts, they are able equally, or rather proportionally, to resist or yield to such external disturbances and blows. For we can demonstrate by geometry that the large machine is not proportionately stronger than the small. Finally, we may say that, for every machine and structure, whether artificial or natural, there is set a necessary limit beyond which neither art nor nature can pass; it is here understood, of course, that the material is the same and the proportion preserved.

SAGR.

My brain already reels. My mind, like a cloud momentarily illuminated by a lightning-flash, is for an instant filled with an unusual light, which now beckons to me and which now suddenly mingles and obscures strange, crude ideas. From what you have said it appears to me impossible to build two similar structures of the same material, but of different sizes and have them proportionately strong; and if this were so, it would not be possible to find two single poles made of the same wood which shall be alike in strength and resistance but unlike in size.

SALV.

So it is, Sagredo. And to make sure that we understand each other, I say that if we take a wooden rod of a certain length and size, fitted, say, into a wall at right angles, i.e., parallel to the horizon, it may be reduced to such a length that it will just support itself; so that if a hair's breadth be added to its length it will break under its own weight and will be the only rod of the kind in the world.* Thus if, for instance, its length be a hundred times its breadth, you will not be able to find another rod whose length is also a hundred times its breadth and which, like the former, is just able to sustain its own weight and no more: all the larger ones will break while all the shorter ones will be strong enough to support something more than their own weight. And this which I have said about the ability to support itself must be understood to apply also to other tests; so that if a piece of scantling[*corrente*] will carry the weight of ten similar to itself, a beam [*trave*] having the same proportions will not be able to support ten similar beams.

Please observe, gentlemen, how facts which at first seem improbable will, even on scant explanation, drop the cloak which has hidden them and stand forth in naked and simple beauty. Who does not know that a horse falling from a height of three or four cubits will break his bones, while a dog falling from the same height or a cat from a height of eight or ten cubits will suffer no injury? Equally harmless would be the fall of a grasshopper from a tower or the fall of an ant from the distance of the moon. Do not children fall with impunity from heights which would cost their elders a broken leg or perhaps a fractured skull? And just as smaller animals are proportionately stronger and more robust than the larger, so also smaller plants are able to stand up better than larger. I am certain you both know that an oak two hundred cubits [*braccia*] high would not be able to sustain its own branches if they were distributed as in a tree of ordinary size; and that nature cannot produce a horse as large as twenty ordinary horses or a giant ten times taller than an ordinary man unless by miracle or by greatly altering the proportions of his limbs and especially of his bones, which would have to be considerably enlarged over the ordinary. Likewise the current belief that, in the case of artificial machines the very large and the small are equally feasible and lasting is a manifest error. Thus, for example, a small obelisk or column or other solid figure can certainly be laid down or set up without danger of breaking, while the very large ones will go to pieces under the slightest provocation, and that purely on account of their own weight. And here I must relate a circumstance which is worthy of your attention as indeed are all events which happen contrary to expectation, especially when a precautionary measure turns out to be a cause of disaster. A large marble column was laid out so that its two ends rested each upon a piece of beam; a little later it occurred to a mechanic that, in order to be doubly sure of its not breaking in the middle by its own weight, it would be wise to lay a third support midway; this seemed to all an excellent idea; but the sequel showed that it was quite the opposite, for not many months passed before the column was found cracked and broken exactly above the new middle support.

SIMP.

A very remarkable and thoroughly unexpected accident, especially if caused by placing that new support in the middle.

SALV.

Surely this is the explanation, and the moment the cause is known our surprise vanishes; for when the two pieces of the column were placed on level ground it was observed that one of the end beams had, after a long while, become decayed and sunken, but that the middle one remained hard and strong, thus causing one half of the column to project in the air without any support. Under these circumstances the body therefore behaved differently from what it would have done if supported only upon the first beams; because no matter how much they might have sunken the column would have gone with them. This is an accident which could not possibly have happened to a small column, even though made of the same stone and having a length corresponding to its thickness, i.e., preserving the ratio between thickness and length found in the large pillar.

SAGR.

I am quite convinced of the facts of the case, but I do not understand why the strength and resistance are not multiplied in the same proportion as the material; and I am the more puzzled because, on the contrary, I have noticed in other cases that the strength and resistance against breaking increase in a larger ratio than the amount of material. Thus, for instance, if two nails be driven into a wall, the one which is twice as big as the other will support not only twice as much weight as the other, but three or four times as much.

SALV.

Indeed you will not be far wrong if you say eight times as much; nor does this phenomenon contradict the other even though in appearance they seem so different.

SAGR.

Will you not then, Salviati, remove these difficulties and clear away these obscurities if possible: for I imagine that this problem of resistance opens up a field of beautiful and useful ideas; and if you are pleased to make this the subject of to-day's discourse you will place Simplicio and me under many obligations.

SALV.

I am at your service if only I can call to mind what I learned from our Academician who had thought much upon this subject and according to his custom had demonstrated everything by geometrical methods so that one might fairly call this a new science. For, although some of his conclusions had been reached by others, first of all by Aristotle, these are not the most beautiful and, what is more important, they had not been proven in a rigid manner from fundamental principles.

...

SALV.

The facts set forth by me up to this point and, in particular, the one which shows that difference of weight, even when very great, is without effect in changing the speed of falling bodies, so that as far as weight is concerned they all fall with equal speed: this idea is, I say, so new, and at first glance so remote from fact, that if we do not have the means of making it just as clear as sunlight, it had better not be mentioned; but having once allowed it to pass my lips I must neglect no experiment or argument to establish it.

SAGR.

Not only this but also many other of your views are so far removed from the commonly accepted opinions and doctrines that if you were to publish them you would stir up a large number of antagonists; for human nature is such that men do not look with favor upon discoveries—either of truth or fallacy—in their own field, when made by others than themselves. They call him an innovator of doctrine, an unpleasant title, by which they hope to cut those knots which they cannot untie, and by subterranean mines they seek to destroy structures which patient artisans have built with customary tools. But as for ourselves who have no such thoughts, the experiments and arguments which you have thus far adduced are fully satisfactory; however if you have any experiments which are more direct or any arguments which are more convincing we will hear them with pleasure.

SALV.

The experiment made to ascertain whether two bodies, differing greatly in weight will fall from a given height with the same speed offers some difficulty; because, if the height is considerable, the retarding effect of the medium, which must be penetrated and thrust aside by the falling body, will be greater in the case of the small momentum of the very light body than in the case of the great force [*violenza*] of the heavy body; so that, in a long distance, the light body will be left behind; if the height be small, one may well doubt whether there is any difference; and if there be a difference it will be inappreciable.

It occurred to me therefore to repeat many times the fall through a small height in such a way that I might accumulate all those small intervals of time that elapse between the arrival of the heavy and light bodies respectively at their common terminus, so that this sum makes an interval of time which is not only observable, but easily observable. In order to employ the slowest speeds possible and thus reduce the change which the resisting medium produces upon the simple effect of gravity it occurred to me to allow the bodies to fall along a plane slightly inclined to the horizontal. For in such a plane, just as well as in a vertical plane, one may discover how bodies of different weight behave: and besides this, I also wished to rid myself of the resistance which might arise from contact of the moving body with the aforesaid inclined plane. Accordingly I took two balls, one of lead and one of cork, the former more than a hundred times heavier than the latter, and suspended them by means of two equal fine threads, each four or five cubits long. Pulling each ball aside from the perpendicular, I let them go at the same instant, and they, falling along the circumferences of circles having these equal strings for semi-diameters, passed beyond the perpendicular and returned along the same path. This free vibration [*per lor medesime le andate e le tornate*] repeated a hundred times showed clearly that the heavy body maintains so nearly the period of the light body that neither in a hundred swings nor even in a thousand will the former anticipate the latter by as much as a single moment [*minimo momento*], so perfectly do they keep step. We can also observe the effect of the medium which, by the resistance which it offers to motion, diminishes the vibration of the cork more than that of the lead, but without altering the frequency of either; even when the arc traversed by the cork did not exceed five or six degrees while that of the lead was fifty or sixty, the swings were performed in equal times.

SIMP.

If this be so, why is not the speed of the lead greater than that of the cork, seeing that the former traverses sixty degrees in the same interval in which the latter covers scarcely six?

SALV.

But what would you say, Simplicio, if both covered their paths in the same time when the cork, drawn aside through thirty degrees, traverses an arc of sixty, while the lead pulled aside only two degrees traverses an arc of four? Would not then the cork be proportionately swifter? And yet such is the experimental fact. But observe this: having pulled aside the pendulum of lead, say through an arc of fifty degrees, and set it free, it swings beyond the perpendicular almost fifty degrees, thus describing an arc of nearly one hundred degrees; on the return swing it describes a little smaller arc; and after a large number of such vibrations it finally comes to rest. Each vibration, whether of ninety, fifty, twenty, ten, or four degrees occupies the same time: accordingly the speed of the moving body keeps on diminishing since in equal intervals of time, it traverses arcs which grow smaller and smaller.

Precisely the same things happen with the pendulum of cork, suspended by a string of equal length, except that a smaller number of vibrations is required to bring it to rest, since on account of its lightness it is less able to overcome the resistance of the air; nevertheless the vibrations, whether large or small, are all performed in time-intervals which are not only equal among themselves, but also equal to the period of the lead pendulum. Hence it is true that, if while the lead is traversing an arc of fifty degrees the cork covers one of only ten, the cork moves more slowly than the lead; but on the other hand it is also true that the cork may cover an arc of fifty while the lead passes over one of only ten or six; thus, at different times, we have now the cork, now the lead, moving more rapidly. But if these same bodies traverse equal arcs in equal times we may rest assured that their speeds are equal.

SIMP.

I hesitate to admit the conclusiveness of this argument because of the confusion which arises from your making both bodies move now rapidly, now slowly and now very slowly, which leaves me in doubt as to whether their velocities are always equal.

SAGR.

Allow me, if you please, Salviati, to say just a few words. Now tell me, Simplicio, whether you admit that one can say with certainty that the speeds of the cork and the lead are equal whenever both, starting from rest at the same moment and descending the same slopes, always traverse equal spaces in equal times?

SIMP.

This can neither be doubted nor gainsaid.

SAGR.

Now it happens, in the case of the pendulums, that each of them traverses now an arc of sixty degrees, now one of fifty, or thirty or ten or eight or four or two, etc.; and when they both swing through an arc of sixty degrees they do so in equal intervals of time; the same thing happens when the arc is fifty degrees or thirty or ten or any other number; and therefore we conclude that the speed of the lead in an arc of sixty degrees is equal to the speed of the cork when the latter also swings through an arc of sixty degrees; in the case of a fifty-degree arc these speeds are also equal to each other; so also in the case of other arcs. But this is not saying that the speed which occurs in an arc of sixty is the same as that which occurs in an arc of fifty; nor is the speed in an arc of fifty equal to that in one of thirty, etc.; but the smaller the arcs, the smaller the speeds; the fact observed is that one and the same moving body requires the same time for traversing a large arc of sixty degrees as for a small arc of fifty or even a very small arc of ten; all these arcs, indeed, are covered in the same interval of time. It is true therefore that the lead and the cork each diminish their speed [*moto*] in proportion as their arcs diminish; but this does not contradict the fact that they maintain equal speeds in equal arcs.

My reason for saying these things has been rather because I wanted to learn whether I had correctly understood Salviati, than because I thought Simplicio had any need of a clearer explanation than that given by Salviati which like everything else of his is extremely lucid, so lucid, indeed, that when he solves questions which are difficult not merely in appearance, but in reality and in fact, he does so with reasons, observations and experiments which are common and familiar to everyone.

In this manner he has, as I have learned from various sources, given occasion to a highly esteemed professor for undervaluing his discoveries on the ground that they are commonplace, and established upon a mean and vulgar basis; as if it were not a most admirable and praiseworthy feature of demonstrative science that it springs from and grows out of principles well-known, understood and conceded by all.

But let us continue with this light diet; and if Simplicio is satisfied to understand and admit that the gravity inherent [*interna gravità*] in various falling bodies has nothing to do with the difference of speed observed among them, and that all bodies, in so far as their speeds depend upon it, would move with the same velocity, pray tell us, Salviati, how you explain the appreciable and evident inequality of motion; please reply also to the objection urged by Simplicio—an objection in which I concur—namely, that a cannon ball falls more rapidly than a bird-shot. From my point of view, one might expect the difference of speed to be small in the case of bodies of the same substance moving through any single medium, whereas the larger ones will descend, during a single pulse-beat, a distance which the smaller ones will not traverse in an hour, or in four, or even in twenty hours; as for instance in the case of stones and fine sand and especially that very fine sand which produces muddy water and which in many hours will not fall through as much as two cubits, a distance which stones not much larger will traverse in a single pulse-beat.

1220 SALV.

The action of the medium in producing a greater retardation upon those bodies which have a less specific gravity has already been explained by showing that they experience a diminution of weight. But to explain how one and the same medium produces such different retardations in bodies which are made of the same material and have the same shape, but differ only in size, requires a discussion more clever than that by which one explains how a more expanded shape or an opposing motion of the medium retards the speed of the moving body. The solution of the present problem lies, I think, in the roughness and porosity which are generally and almost necessarily found in the surfaces of solid bodies. When the body is in motion these rough places strike the air or other ambient medium. The evidence for this is found in the humming which accompanies the rapid motion of a body through air, even when that body is as round as possible. One hears not only humming, but also hissing and whistling, whenever there is any appreciable cavity or elevation upon the body. We observe also that a round solid body rotating in a lathe produces a current of air. But what more do we need? When a top spins on the ground at its greatest speed do we not hear a distinct buzzing of high pitch? This sibilant note diminishes in pitch as the speed of rotation slackens, which is evidence that these small rugosities on the surface meet resistance in the air. There can be no doubt, therefore, that in the motion of falling bodies these rugosities strike the surrounding fluid and retard the speed; and this they do so much the more in proportion as the surface is larger, which is the case of small bodies as compared with greater.

SIMP.

Stop a moment please, I am getting confused. For although I understand and admit that friction of the medium upon the surface of the body retards its motion and that, if other things are the same, the larger surface suffers greater retardation, I do not see on what ground you say that the surface of the smaller body is larger. Besides if, as you say, the larger surface suffers greater retardation the larger solid should move more slowly, which is not the fact. But this objection can be easily met by saying that, although the larger body has a larger surface, it has also a greater weight, in comparison with which the resistance of the larger surface is no more than the resistance of the small surface in comparison with its smaller weight; so that the speed of the larger solid does not become less. I therefore see no reason for expecting any difference of speed so long as the driving weight [*gravità movente*] diminishes in the same proportion as the retarding power [*facoltà ritardante*] of the surface.

1255 SALV.

I shall answer all your objections at once. You will admit, of course, Simplicio, that if one takes two equal bodies, of the same material and same figure, bodies which would therefore fall with equal speeds, and if he diminishes the weight of one of them in the same proportion as its surface (maintaining the similarity of shape) he would not
1260 thereby diminish the speed of this body.

SIMP.

This inference seems to be in harmony with your theory which states that the weight of a body has no effect in either accelerating or retarding its motion.

SALV.

1265 I quite agree with you in this opinion from which it appears to follow that, if the weight of a body is diminished in greater proportion than its surface, the motion is retarded to a certain extent; and this retardation is greater and greater in proportion as the diminution of weight exceeds that of the surface.

SIMP.

1270 This I admit without hesitation.

SALV.

Now you must know, Simplicio, that it is not possible to diminish the surface of a solid body in the same ratio as the weight, and at the same time maintain similarity of figure. For since it is clear that in the case of a diminishing solid the weight grows less in
1275 proportion to the volume, and since the volume always diminishes more rapidly than the surface, when the same shape is maintained, the weight must therefore diminish more rapidly than the surface. But geometry teaches us that, in the case of similar solids, the ratio of two volumes is greater than the ratio of their surfaces; which, for the sake of better understanding, I shall illustrate by a particular case.

1280 Take, for example, a cube two inches on a side so that each face has an area of four square inches and the total area, i. e., the sum of the six faces, amounts to twenty-four square inches; now imagine this cube to be sawed through three times so as to divide it into eight smaller cubes, each one inch on the side, each face one inch square, and the total surface of each cube six square inches instead of twenty-four as in the case of the larger cube. It is evident therefore that the surface of the little cube is only one-fourth that of the larger, namely, the ratio of six to twenty-four; but the volume of the solid cube itself is only one-eighth; the volume, and hence also the weight, diminishes therefore much more rapidly than the surface. If we again divide the little cube into eight others we shall have, for the total surface of one of these, one and one-half square inches, which is one-sixteenth of the surface of the original cube; but its volume is only one-sixty-fourth part. Thus, by two divisions, you see that the volume is diminished four times as much as the surface. And, if the subdivision be continued until the original solid be reduced to a fine powder, we shall find that the weight of one of these smallest particles has diminished hundreds and hundreds of times as much as its surface. And this which I have illustrated in the case of cubes holds also in the case of all similar solids, where the volumes stand in sesquialteral ratio to their surfaces. Observe then how much greater the resistance, arising from contact of the surface of the moving body with the medium, in the case of small bodies than in the case of large; and when one considers that the rugosities on the very small surfaces of fine dust particles are perhaps no smaller than those on the surfaces of larger solids which have been carefully polished, he will see how important it is that the medium should be very fluid and offer no resistance to being thrust aside, easily yielding to a small force. You see, therefore, Simplicio, that I was not mistaken when, not long ago, I said that the surface of a small solid is comparatively greater than that of a large one.

SIMP.

I am quite convinced; and, believe me, if I were again beginning my studies, I should follow the advice of Plato and start with mathematics, a science which proceeds very cautiously and admits nothing as established until it has been rigidly demonstrated.

Third Day: Change of Position. [*De Motu Locali*]

MY purpose is to set forth a very new science dealing with a very ancient subject. There is, in nature, perhaps nothing older than motion, concerning which the books written by philosophers are neither few nor small; nevertheless I have discovered by experiment some properties of it which are worth knowing and which have not hitherto been either observed or demonstrated. Some superficial observations have been made, as, for instance, that the free motion [*naturalem motum*] of a heavy falling body is continuously accelerated; but to just what extent this acceleration occurs has not yet been announced; for so far as I know, no one has yet pointed out that the distances traversed, during equal intervals of time, by a body falling from rest, stand to one another in the same ratio as the odd numbers beginning with unity.

It has been observed that missiles and projectiles describe a curved path of some sort; however no one has pointed out the fact that this path is a parabola. But this and other facts, not few in number or less worth knowing, I have succeeded in proving; and what I consider more important, there have been opened up to this vast and most excellent science, of which my work is merely the beginning, ways and means by which other minds more acute than mine will explore its remote corners.

This discussion is divided into three parts; the first part deals with motion which is steady or uniform; the second treats of motion as we find it accelerated in nature; the third deals with the so-called violent motions and with projectiles.

UNIFORM MOTION

In dealing with steady or uniform motion, we need a single definition which I give as follows:

Definition

By steady or uniform motion, I mean one in which the distances traversed by the moving particle during any equal intervals of time, are themselves equal.

Caution

We must add to the old definition (which defined steady motion simply as one in which equal distances are traversed in equal times) the word "any," meaning by this, all equal intervals of time; for it may happen that the moving body will traverse equal distances during some equal intervals of time and yet the distances traversed during some small portion of these time-intervals may not be equal, even though the time-intervals be equal.

From the above definition, four axioms follow, namely:

Axiom I

In the case of one and the same uniform motion, the distance traversed during a longer interval of time is greater than the distance traversed during a shorter interval of time.

Axiom II

1345 In the case of one and the same uniform motion, the time required to traverse a greater distance is longer than the time required for a less distance.

Axiom III

In one and the same interval of time, the distance traversed at a greater speed is larger than the distance traversed at a less speed.

1350 **Axiom IV**

The speed required to traverse a longer distance is greater than that required to traverse a shorter distance during the same time-interval.

NATURALLY ACCELERATED MOTION

The properties belonging to uniform motion have been discussed in the preceding section; but
1355 accelerated motion remains to be considered.

And first of all it seems desirable to find and explain a definition best fitting natural phenomena. For anyone may invent an arbitrary type of motion and discuss its properties; thus, for instance, some have imagined helices and conchoids as described by certain motions which are not met with in nature, and have very commendably established the properties which these curves possess in
1360 virtue of their definitions; but we have decided to consider the phenomena of bodies falling with an acceleration such as actually occurs in nature and to make this definition of accelerated motion exhibit the essential features of observed accelerated motions. And this, at last, after repeated efforts we trust we have succeeded in doing. In this belief we are confirmed mainly by the consideration that experimental results are seen to agree with and exactly correspond with those properties which
1365 have been, one after another, demonstrated by us. Finally, in the investigation of naturally accelerated motion we were led, by hand as it were, in following the habit and custom of nature herself, in all her various other processes, to employ only those means which are most common, simple and easy. For I think no one believes that swimming or flying can be accomplished in a manner simpler or easier than that instinctively employed by fishes and birds.

1370 When, therefore, I observe a stone initially at rest falling from an elevated position and continually acquiring new increments of speed, why should I not believe that such increases take place in a manner which is exceedingly simple and rather obvious to everybody? If now we examine the matter carefully we find no addition or increment more simple than that which repeats itself always in the same manner. This we readily understand when we consider the intimate relationship between time and motion; for just as uniformity of motion is defined by and conceived through equal times and equal spaces (thus we call a motion uniform when equal distances are traversed during equal time-intervals), so also we may, in a similar manner, through equal time-intervals, conceive additions of speed as taking place without complication; thus we may picture to our mind a motion as uniformly and continuously accelerated when, during any equal intervals of time whatever, equal increments of speed are given to it. Thus if any equal intervals of time whatever have elapsed, counting from the time at which the moving body left its position of rest and began to descend, the amount of speed acquired during the first two time-intervals will be double that acquired during the first time-interval alone; so the amount added during three of these time-intervals will be treble; and that in four, quadruple that of the first time-interval. To put the matter more clearly, if a body were to continue its motion with the same speed which it had acquired during the first time-interval and were to retain this same uniform speed, then its motion would be twice as slow as that which it would have if its velocity had been acquired during *two* time-intervals.

And thus, it seems, we shall not be far wrong if we put the increment of speed as proportional to the increment of time; hence the definition of motion which we are about to discuss may be stated as follows: A motion is said to be uniformly accelerated, when starting from rest, it acquires, during equal time-intervals, equal increments of speed.

Reading 5
Descartes: Le Monde

Translated by Jonathan English

Chapter 1

In proposing to treat here about light, the first thing of which I want to inform you is that there can be a difference between the sensation that we have of light, that is to say the idea that forms in our imagination by the mediation of our eyes, and that which is in the objects which produce in us this sensation, that is to say that which is in the flame or in the sun, which is called light. For even though we can normally convince ourselves that the ideas we have in our thoughts entirely resemble the objects from which they proceed, nevertheless I do not at all see any reason that this must be granted; but I say there are, on the contrary, some experiences which make us doubt it.

You well know that words, not having any resemblance to the things they signify, do not prevent us from conceiving of them, and often even aside from us noticing the sound of the word, or their syllables. It may easily happen that after having heard a conversation, of which we shall have understood the sense of it very well, we will not be able to say in which language it was spoken. But, it these words, which signify nothing but by the institution of men, suffice for us to conceive of these things, with which they do not have any resemblance, then why is nature herself unable to have also established certain signs, which make us have sensation of light, even though this sign may be nothing in itself, which resembles this sensation? And is it not for this that it has established laughing and crying in order for us to read joy and sorrow on the faces of men?

But you say, perhaps, that our ears only truly sense the sound of the words, and our eyes only the countenance of him who laughs or cries, and that which is our mind, which having retained that which is signified by these words and this countenance, represents it to us at the same time. To this I might say that this is our mind all the same, which represents the idea of the light, every time that the action which signifies it touches our eye. But without losing the time to dispute it, I ought rather to bring to bear another example.

Think you then that we do not take notice of the significance of the words, and that we hear only their sound; that the idea of this sound, which forms in our thoughts, must be each thing of sensation which is the cause of the idea? One man opens his mouth, stirs his tongue, pushes out his breath: I see nothing in all these actions which may not be more different from the idea of the sound than they have us imagine. And most of the philosophers assure that the sound is not a thing other than the quivering of the air, which comes to beat our ears; even more so, if the sense of the eye reports to our thoughts the true image of its objects, it must, instead of us conceiving the sound, should we be made to conceive of the movement of parts of the air which quiver against our ears. But, because no everyone may be able to want to believe that which the philosophers say, I will bring still another example.

Touch is that of all our senses that one esteems the least false and the most assured; thus it is the stronger if I show you that even touch makes us conceive many ideas that do that resemble in any fashion the objects that produce them, I do think that would ought to find it strange, if I say that sight may be made the same. Now, there is no one who does not know that the ideas of tickling and of pain, which form themselves in our thoughts at the occasion of the bodies touching us from without, do not have any resemblance to these things. One gently passes a feather across the lips of a sleeping child and he feels like someone is tickling him: do you think that the idea of tickling, which he conceives, resembles such a thing as that which is in this feather? A knight returns from combat: while in the heat of battle, he may be able to be wounded without perceiving it; but as soon as he begins to cool down, feels some pain, he believes himself wounded: someone calls for a surgeon, another removes his armor, they search him and find that a buckle or strap, which was caught under his armor, pressed and bothered him. If its touch, in making him sense this strap, had impressed the image in his thoughts, he would not need to call a surgeon because of what he sensed.

Now, I do not at all see a reason which obliges us to believe that that which is in the objects from which come the sensation of light to us, must more resemble this sensation than the actions of a feather and a strap to the sense of tickling or to pain. But all this time, I have not at all brought these examples in order to make you believe absolutely that this light is different in the objects and in our eyes; but only that you may finally question it, and that guarding yourself against being preoccupied to the contrary, you will be able soon to search out with me that which it is.

Chapter 2

I only know two sorts of bodies in the world in which light is found, namely the stares and flame, or fire. And because the stars are without a doubt farther from the knowledge of men than is the fire or flame, I will test first explaining my remarks touching flame.

When fire burns some wood, or some other similar material, we are able to see with the eye that it moves the small parts of this wood, and separates the from the other, also transforming the most subtle parts into fire, air and smoke, and leaving the largest parts in cinders. Someone else may thus imagine, if he wants, in this wood the form of "Fire", the quality of "Heat", and the action of "Burning", and all such diverse things; for myself, who fears finding in myself that I am supposing something there more than that which I see necessarily ought to be there, I content myself with conceiving there the moving of its parts. For, put the "Fire" in there, put in the "Heat", and let it "Burn", as much as you please, if you do not at all suppose with this that it has there any of these parts which move or which detach from their neighbors, I cannot imagine that it receives any alteration or change. And, on the contrary, remove from it the "Fire" and the "Heat, and prevent it from "Burning": provided only that you accord me that there is some power which violently moves the most subtle of the parts, and which separates them from the greater parts, I find that this alone is able to perform in it all the same changes that on experiences when it burns.

But, as it does not seem to be possible to think that a body is able to move another if it is not itself moving at the same time, I conclude from this that the body of the fame which acts against the wood is composed of small parts which move themselves separately one from another in a very quick and violent movement, and which moving themselves forcefully like this, push and move with themselves the parts of the bodies that they touch that do not have too much resistance to them. I say that its parts move themselves separately from one another: for while many often work in accord and conspire together in order to produce the same effect, we see all the time that each of them acts in its particular way against the bodies that they touch. I say also that their movement is very quick and violent: for, being such small parts that our sight cannot distinguish them, they would not have enough force to act against the other bodies, so the quickness of their movement compensates for their lack of size.

I do not mention the direction that each one moves: for if you consider that the ability to move itself and that which determines from where the movement comes are two different things and which are able to be the one without the other (this I have explained in the Dioptics), you judge easily that each moves itself easily in the way in which it is given the least difficulty by the disposition of the bodies which surround it, and which in the same flame, it is able there to have parts which rise and others which sink, all straight or in a ring or in every direction without changing nothing of the flame's nature. Even more so, if you see them ascend almost every time, one must not think that this must be for another reason, than because the other bodies which touch them are found to be disposed to make more resistance in all the other directions.

But after having recognized that the parts of the flame stir themselves in this manner and that it is sufficient to this of these movement in order to comprehend how it is able to consume wood and to burn, let us examine, I pray you, is the same is not also sufficient in order to make us also comprehend, how it warms us and gives us light. For if it is found to be thus, it will not be necessary that there is in flame any other quality, and we will be able to say that it is movement alone that, according to the different effects that it produces, is sometimes called "heat" and at other times "light".

Now, for that which is heat, the sense that we have of it is able, it seems to me, to be taken for a species of pain when it is violent, and sometimes for a species of tickling when it is moderate. And as we have already said that that there is nothing except our thought which may resemble the ideas that we conceive of tickling and of pain, we are able to well believe also that there is nothing which may resemble that which we conceive of Heat, but that all that which is able to move diversely the small parts of our hands, or from some other point in our body, is able to excite this sensation in us. Many similar experiences promote this opinion, for, in rubbing the hands alone, they become warm, and all other bodies may also be warmed without being placed in the flame, provided only that it may be agitated and shaken in such a strong manner that many of these small parts are moved, and are able to move with them those of our hands.

For that which is Light, one is also able to well conceive that the same movement that is in the flame suffices for us to have sensation of light. But because it is in this that the principle part of my design consists, I want to try to explain it at great length and take up my discourse from the beginning.

Chapter 7

But I do not want to delay any longer telling you by what means Nature alone is able to untangle the confusion of Chaos of which I have spoken, and which are the Laws that God has imposed on them.

Know this, first, that by Nature I do not at all mean here some "Goddess", or some other kind of imaginary potency, but that I use that word in order to signify Matter itself, such as that I consider it with all the qualities taken all together that I have attributed to it, and under the condition that God continues to conserve it in that same fashion that he has created it. For from this alone, that he that he continues also to conserve it, it follows, by necessity, that there must be many changes in its parts, which not being able, it seems to me, to be properly attributed to that action of God, because it does not change at all, I attribute to Nature, and the rules to which the changes follow, I name the Law of Nature.

In order to better understand this, remember that between the qualities of matter, we have supposed that these parts have had diverse movements since the beginning when they had been created, and besides that, they completely touch each other on all sides without there being any void between them. From whence it follows by necessity, that from that moment on, in beginning to move, they have commenced also to change and diversify their movements by the encounters between each other, and also that if God conserves them after the same fashion that he had created them, he does not conserve them in the same state. That is to say, God acting always the same, and by consequence produces always the same effect in the substance, it is found, as by accident, many diversities in this effect. And it is easy to believe that God, who as everyone must know, is immutable, acts always in the same fashion. But without engaging myself more in these Metaphysical consideration, I lay out here two or three principle rules, following from which it must be thought that God causes the Nature of this new world to act, and which will suffice, as I believe, in order to cause you to believe all the others.

The first is this: That each part of matter in particular continues always being in the same state, while the meeting with others does not at all keep it from changing. That is to say, if it has such a size, it will never become smaller unless the others divide it; if it is round or square, it will never change its figure without the others acting against it; if it is stopped in some place, it will never depart from there unless some others expel it; and if it has at one time commenced moving, it will always continue moving with an equal force until some others stop or slow it.

There is not a person who does not believe that the same rule observed in the ancient world, touching the size, figure, position, and a thousand other similar things; but the Philosophers have exempted movement, which is however the thing that I most expressly desire to comprehend here. And do not think that what I have designed is to contradict them: the movement of which they speak is so different a sort than that which I conceive that it may easily happen that that which is true in the one may not be so in the other.

They avow themselves that their nature is of a kind little known; and in order to render it in some intelligible fashion, they as yet have not explained it more clearly than these terms: *Motus est actus entis in potentia, prout in potential est*, which are for me so obscure, that I am constrained to leave them here in their language, because I do not know how to interpret them. (And in effect these words: "Movement is the action of a being in potency, as that it is in potency", are not more clear even rendered in [English].) But, on the contrary, the nature of movement which I intend here to speak, is so easy a connection, that even the Geometers, who among all men are the most studied at conceiving very distinctly the things that they consider, have judged it more simple and more intelligible than those of their surfaces and their lines. Thus it appears in that they explain the line by the movement of a point, and the surface by that of a line.

The philosopher also suppose several movements that they think can be made without the body changing place as that which they call *Motus ad formam, motus ad calorem, motus ad quantitatem* (movement of the form, movement of the heat, movement of the quantity) and a thousand others. And myself, I do not know any than that which is easier to conceive than the lines of Geometry, which make the bodies push into the place of another and occupy successively all the spaces which are between them.

Other than that, they attribute to the least of the movements a being much more solid and more veritable than they have of rest, which they say is only a privation of movement. And for myself, I believe that rest is also well a quality, which must be attributed to the material, while it resides in a place, like its movement is a quality which is attributed to it, while it is changing place.

Finally the movement of which they speak is of a nature so strange that instead of all the other things that have for the end their perfection, and try only to preserve themselves, it has no other end goal than rest, and against all the Laws of Nature, it tries itself to destroy itself. But, on the contrary, those which I suppose are the same Laws of Nature that are generally all the dispositions and all the qualities which are found in the matter; also those which the scholars call *Modos et entia rationis cum fundamento in re* (modes and being of reason with the foundation in the thing), like *Qualitates reales* (their real qualities), in which I confess ingenuously to not find more reality than in the others.

I suppose for a second rule: that when one body pushes another, it cannot give the other any movement, unless it loses at the same time some of its own. This rule joined with the preceding reports well with all experiences, in each we see that a body commences or ceases to move, because it is pushed or stopped by some other. For, having supposed the preceding, we are exempt from the sentence where the scholars find themselves, when they want to render the reason why one stone continues to move for some time after being out of the hand which has thrown it. For one may rather demand of us, why does it not continue to always move? But the reason is easy to give. For who is able to deny that air, in which it moves, does not offer some resistance? One hears it whistle, then it is divided, and if one moves it with a fan, or some other very lightweight and very elongated body, one is able to sense the same by the weight of the hand that it impedes the movement, very far from continuing it, thus someone had wanted to say. But if the one lacks the explanation of the effect of the resistance following our second law, and that one thinks that, the more a body is able to resist, the more it is able to stop the movement of others, thus that may be first one is able to persuade. One will have again the sentence of reason rendered, why the movement of this stone amortizes rather in meeting a soft body and in which the resistance is mediocre, than when it meets one more durable and which resists it more? Likewise, why does it as soon as it has met a little effort at last, return immediately upon its own trail, rather than stopping or interrupting its movement for its subject? Instead, suppose this Law, it is not at all in this difficulty, for it shows us that the movement of a body is not slowed by the meeting of another in proportion of how much it resists it, but only in proportion of how much its resistance is overcome, and in obeying the law, it receives in time the force of moving that the other left.

Now, even though in most of the movements that we see in the true world, we are not able to perceive that the bodies which commence or cease movement, may be pushed or stopped by some other: we have not judged on this occasion that these two rules may not be exactly observed. For it is certain that these bodies are able often to receive their agitation from the two elements of Air and Fire, which find themselves always among them, without being sensibly present, thus as has sometime been said, or even from the grosser air, which is not able to be sensed; and that they are able to transfer it, sometimes to the grosser air, and sometimes to all the mass of the earth, in which being dispersed, it is not thus able to be perceived.

But even if all that which our senses had ever experienced in the true world seemed manifestly to be contrary to that which is contained in these two rules, the reason which has taught them to me them seems to me so strong, that I will not leave them to the belief, being obliged to suppose them in the news that I say to you. For what foundation more firm and more solid is one able to find in order to establish a truth, even though on wants to choose to wish it, than to take even the constancy and immutability which is in God?

Now it is that these two rules follow manifestly from this alone, that God is immutable and that acting always in the same way, he produces always the same effect. For, supposing that he has placed a certain quantity of movement in all the matter in general, from the first instant that he had created it, it must be avowed that it conserves there always that amount, or not to be believed that he acts always in the same way. And supposing with this that from the first instant the diverse parts of matter in which these movements are found unequally dispersed, has commenced to retain them, or to transfer them one to another according in so much as they have been able to have the force, it must necessarily be thought that he makes them always continue in the same thing. And this is that which these two rules contain.

I will add as a third rule: That when a body moves, even though its movement often travels along a curved line, and though it is never able to make one which is in some way not circular, as has been said previously, however each of its particular parts tend always to continue themselves in a straight line. And thus their action, that is to say the inclination that they have to move, is in different from their movement.

For example, if one turns a wheel on its axle, even though all its parts go round, since being joined the one to the other always their inclination is to go straight, as is made clear if by accident some part detaches from the others. For as soon as it is free, it continues in a straight line.

In the same way, when one turns a stone in a sling, not only does it go always straight as soon as it is released, but while at the time that it is there, it presses the middle of the sling and pulling the cord taut showing evidently by this that it always inclines to go in a straight line and that it only goes round by constraint.

This rule rests on the same foundation as that of the two others and only depends on how God conserves each thing by a continuous action and by consequence the he does not conserve it at all such that it may have been some time before, but precisely such as it is in the same instant that he conserves it. Now it is that of these movements, there is only the straight which is entirely simple and which all nature may be comprised of in any instant. For, in order to conceive it, it suffices to think that a body is in action in order to move towards a certain direction that which is found in each of the instants which are able to be determined while it moves. Instead, in order to conceive the circular movement, or some other possible movement, it must be at least considered in two instants or some two of its parts and the connection between them.

But so that the philosophers, or more so the sophists, do not take here the occasion to exercise their subtle superfluity, remark that I did not say that the straight movement is able to be made in an instant; but only that all that which is required to produce it, is found in the bodies in each instant which is able to be determined while they move and not all that which is require for the production of circular motion.

Reading 6
Newton: Principia: Definitions

Translation by Andrew Motte

THE AUTHOR S PREFACE

SINCE the ancients (as we are told by Pappus), made great account of the science of mechanics in the investigation of natural things: and the moderns, laying aside substantial forms and occult qualities, have endeavoured to subject the phenomena of nature to the laws of mathematics, I have in this treatise cultivated mathematics so far as it regards philosophy. The ancients considered mechanics in a twofold respect; as rational, which proceeds accurately by demonstration; and practical. To practical mechanics all the manual arts belong, from which mechanics took its name. But as artificers do not work with perfect accuracy, it comes to pass that mechanics is so distinguished from geometry, that what is perfectly accurate is called geometrical, what is less so, is called mechanical. But the errors are not in the art, but in the artificers. He that works with less accuracy is an imperfect mechanic; and if any could work with perfect accuracy, he would be the most perfect mechanic of all; for the description if right lines and circles, upon which geometry is founded, belongs to mechanics. Geometry does not teach us to draw these lines, but requires them to be drawn; for it requires that the learner should first be taught to describe these accurately, before he enters upon geometry; then it shows how by these operations problems may be solved. To describe right lines and circles are problems, but not geometrical problems. The solution of these problems is required from mechanics; and by geometry the use of them, when so solved, is shown; and it is the glory of geometry that from those few principles, brought from without, it is able to produce so many things. Therefore geometry is founded in mechanical practice, and is nothing but that part of universal mechanics which accurately proposes and demonstrates the art of measuring. But since the manual arts are chiefly conversant in the moving of bodies, it comes to pass that geometry is commonly referred to their magnitudes, and mechanics to their motion. In this sense rational mechanics will be the science of motions resulting from any forces whatsoever, and of the forces required to produce any motions, accurately proposed and demonstrated. This part of mechanics was cultivated by the ancients in the five powers which relate to manual arts, who considered gravity (it not being a manual power), ho Otherwise than as it moved weights by those powers. Our design not respecting arts, but philosophy, and our subject not manual but natural powers, we consider chiefly those things which relate to gravity, levity, elastic force, the resistance of fluids, and the like forces, whether attractive or impulsive; and therefore we offer this work as the mathematical principles of philosophy; for all the difficulty of philosophy seems to consist in this from the phenomena of motions to investigate the forces of nature, and then from these forces to demonstrate the other phenomena; and to this end the general propositions in the first and second book are directed. In the third book we give an example of this in the explication of the System of the World: for by the propositions mathematically demonstrated in the former books, we in the third derive from the celestial phenomena the forces of gravity with which bodies tend to the sun and the several planets. Then from these forces, by other propositions which are also mathematical, we deduce the motions of the planets, the comets, the moon, and the sea. I wish we could derive the rest of the phenomena of nature by the same kind of reasoning from mechanical principles; for I am induced by many reasons to suspect that they may all depend upon certain forces by which the particles of bodies by some causes hitherto unknown, are either mutually impelled towards each other, and cohere in regular figures, or are repelled and recede from each other; which forces being unknown, philosophers have hitherto attempted the search of nature in vain ; but I hope the principles here laid down will afford some light either to this or some truer method of philosophy. In the publication of this work the most acute and universally learned Mr. Edmund Halley not only assisted me with his pains in correcting the press and taking care of the schemes, but it was to his solicitations that its becoming public is owing; for when he had obtained of me my demonstrations of the figure of the celestial orbits, he continually pressed me to communicate the same to the Royal Society, who afterwards, by their kind encouragement and entreaties, engaged me to think of publishing them. But after I had begun to consider the inequalities of the lunar motions, and had entered upon some other things relating to the laws and measures of gravity, and other forces: and the figures that would be described by bodies attracted according to given laws; and the motion of several bodies moving among themselves; the motion of bodies in resisting mediums; the forces, densities, and motions, of mediums; the orbits of the comets, and such like; deferred that publication till I had made a search into those matters, and could put forth the whole together. What relates to the lunar motions (being imperfect), I have put all together in the corollaries of Prop. 66, to avoid being obliged to propose and distinctly demonstrate the several things there contained in a method more prolix than the subject deserved, and interrupt the series of the several propositions. Some things, found out after the rest, I chose to insert in places less suitable, rather than change the number of the propositions and the citations. I heartily beg that what I have here done may be read with candour; and that the defects in a subject so difficult be not so much reprehended as kindly supplied, and investigated by new endeavours of my readers.

ISAAC NEWTON.

DEFINITIONS.

DEFINITION I.

The quantity of matter is the measure of the same, arising from its density and Bulk conjunctly.

THUS air of a double density, in a double space, is quadruple in quantity; in a triple space, sextuple in quantity. The same thing is to be understood of snow, and fine dust or powders, that are condensed by compression or liquefaction and of all bodies that are by any causes whatever differently condensed. I have no regard in this place to a medium, if any such there is, that freely pervades the interstices between the parts of bodies. It is this quantity that I mean hereafter everywhere under the name of body or mass. And the same is known by the weight of each body; for it is proportional to the weight, as I have found by experiments on pendulums, very accurately made, which shall be shewn hereafter.

DEFINITION II.

The quantity of motion is the measure of the same, arising from the velocity and quantity of matter conjunctly.

The motion of the whole is the sum of the motions of all the parts; and therefore in a body double in quantity, with equal velocity, the motion is double; with twice the velocity, it is quadruple.

DEFINITION III.

The vis insita, *or innate force of matter, is a power of resisting, by which every body, as much as in it lies, endeavours to persevere in its present state, whether it be of rest, or of moving uniformly forward in a right line.*

This force is ever proportional to the body whose force it is; and differs nothing from the inactivity of the mass, but in our manner of conceiving it. A body, from the inactivity of matter, is not without difficulty put out of its state of rest or motion. Upon which account, this *vis insita*, may, by a most significant name, be called *vis inertia*, or force of inactivity. But a body exerts this force only, when another force, impressed upon it, endeavours to change its condition; and the exercise of this force may be considered both as resistance and impulse; it is resistance, in so far as the body, for maintaining its present state, withstands the force impressed; it is impulse, in so far as the body, by not easily giving way to the impressed force of another, endeavours to change the state of that other. Resistance is usually ascribed to bodies at rest, and impulse to those in motion; but motion and rest, as commonly conceived, are only relatively distinguished; nor are those bodies always truly at rest, which commonly are taken to be so.

DEFINITION IV.

An impressed force is an action exerted upon a body, in order to change its state, either of rest, or of moving uniformly forward in a right line.

This force consists in the action only; and remains no longer in the body, when the action is over. For a body maintains every new state it acquires, by its *vis inertiae* only. Impressed forces are of different origins as from percussion, from pressure, from centripetal force.

DEFINITION V.

A centripetal force is that by which bodies are drawn or impelled, or any way tend, towards a point as to a centre.

Of this sort is gravity, by which bodies tend to the centre of the earth magnetism, by which iron tends to the loadstone ; and that force, whatever it is, by which the planets are perpetually drawn aside from the rectilinear motions, which otherwise they would pursue, and made to revolve in curvilinear orbits. A stone, whirled about in a sling, endeavours to recede from the hand that turns it; and by that endeavour, distends the sling, and that with so much the greater force, as it is revolved with the greater velocity, and as soon as ever it is let go, flies away. That force which opposes itself to this endeavour, and by which the sling perpetually draws back the stone towards the hand, and retains it in its orbit, because it is directed to the hand as the centre of the orbit, I call the centripetal force. And the same thing is to be understood of all bodies, revolved in any orbits. They all endeavour to recede from the centres of their orbits; and were it not for the opposition of a contrary force which restrains them to, and detains them in their orbits, which I therefore call centripetal, would fly off in right lines, with an uniform motion. A projectile, if it was not for the force of gravity, would not deviate towards the earth, tut would go off from it in a right line, and that with an uniform motion, if the resistance of the air was taken away. It is by its gravity that it is drawn aside perpetually from its rectilinear course, and made to deviate towards the earth, more or less, according to the force of its gravity, and the velocity of its motion. The less its gravity is, for the quantity of its matter, or the greater the velocity with which it is projected, the less will it deviate from a rectilinear course, and the farther it will go. If a leaden ball, projected from the top of a mountain by the force of gunpowder with a given velocity, and in a direction parallel to the horizon, is carried in a curved line to the distance of two miles before it falls to the ground; the same, if the resistance of the air were taken away, with a double or decuple velocity, would fly twice or ten times as far. And by increasing the velocity, we may at pleasure increase the distance to which it might be projected, and diminish the curvature of the line, which it might describe, till at last it should fall at the distance of 10, 30, or 90 degrees, or even might go quite round the whole earth before it falls; or lastly, so that it might never fall to the earth, but go forward into the celestial spaces, and proceed in its motion *in infinitum*. And after the same manner that a projectile, by the force of gravity, may be made to revolve in an orbit, and go round the whole earth, the moon also, either by the force of gravity, if it is endued with gravity, or by any other force, that impels it towards the earth, may be perpetually drawn aside towards the earth, out of the rectilinear way, which by its innate force it would pursue; and would be made to revolve in the orbit which it now describes ; nor could the moon without some such force, be retained in its orbit. If this force was too small, it would not sufficiently turn the moon out of a rectilinear course: if it was too great, it would turn it too much, arid draw down the moon from its orbit towards the earth. It is necessary, that the force be of a just quantity, and it belongs to the mathematicians to find the force, that may serve exactly to retain a body in a given orbit, with a given velocity; and vice versa, to determine the curvilinear way, into which a body projected from a given place, with a given velocity, may be made to deviate from its natural rectilinear way, by means of a given force.

The quantity of any centripetal force may be considered as of three kinds; absolute, accelerative, and motive.

DEFINITION VI.

The absolute quantity of a centripetal force is the measure of the same proportional to the efficacy of the cause that propagates it from the centre, through the spaces round about.

Thus the magnetic force is greater in one load-stone and less in another according to their sizes and strength of intensity.

DEFINITION VII.

The accelerative quantity of a centripetal force is the measure of the same, proportional to the velocity which it generates in a given time.

Thus the force of the same load-stone is greater at a less distance, and less at a greater: also the force of gravity is greater in valleys, less on tops of exceeding high mountains; and yet less (as shall hereafter be shown), at greater distances from the body of the earth; but at equal distances, it is the same everywhere; because (taking away, or allowing for, the resistance of the air), it equally accelerates all falling bodies, whether heavy or light, great or small.

DEFINITION VIII.

The motive quantity of a centripetal force, is the measure of the same proportional to the motion which it generates in a given time.

Thus the weight is greater in a greater body, less in a less body; and. in the same body, it is greater near to the earth, and less at remoter distances. This sort of quantity is the centripetency, or propension of the whole body towards the centre, or, as I may say, its weight; and it is always known by the quantity of an equal and contrary force just sufficient to hinder the descent of the body.

These quantities of forces, we may, for brevity's sake, call by the names of motive, accelerative, and absolute forces; and, for distinction's sake, consider them, with respect to the bodies that tend to the centre; to the places of those bodies; and to the centre of force towards which they tend; that is to say, I refer the motive force to the body as an endeavour and propensity of the whole towards a centre, arising from the propensities of the several parts taken together ; the accelerative force to the place of the body, as a certain power or energy diffused from the centre to all places around to move the bodies that are in them: and the absolute force to the centre, as endued with some cause, without which those motive forces would not be propagated through the spaces round about ; whether that cause be some central body (such as is the load-stone, in the centre of the magnetic force, or the earth in the centre of the gravitating force), or anything else that does not yet appear. For I here design only to give a mathematical notion of those forces, without considering their physical causes and seats.

Wherefore the accelerative force will stand in the same relation to the motive, as celerity does to motion. For the quantity of motion arises from the celerity drawn into the quantity of matter: and the motive force arises from the accelerative force drawn into the same quantity of matter. For the sum of the actions of the accelerative force, upon the several; articles of the body, is the motive force of the whole. Hence it is, that near the surface of the earth, where the accelerative gravity, or force productive of gravity, in all bodies is the same, the motive gravity or the weight is as the body: but if we should ascend to higher regions, where the accelerative gravity is less, the weight would be equally diminished, and would always be as the product of the body, by the accelerative gravity. So in those regions, where the accelerative gravity is diminished into one half, the weight of a body two or three times less, will be four or six times less.

I likewise call attractions and impulses, in the same sense, accelerative, and motive; and use the words attraction, impulse or propensity of any sort towards a centre, promiscuously, and indifferently, one for another; considering those forces not physically, but mathematically: wherefore, the reader is not to imagine, that by those words, I anywhere take upon me to define the kind, or the manner of any action, the causes or the physical reason thereof, or that I attribute forces, in a true and physical sense, to certain centres (which are only mathematical points) ; when at any time I happen to speak of centres as attracting, or as endued with attractive powers.

SCHOLIUM.

Hitherto I have laid down the definitions of such words as are less known, and explained the sense in which I would have them to be understood in the following discourse. I do not define time, space, place and motion, as being well known to all. Only I must observe, that the vulgar conceive those quantities under no other notions but from the relation they bear to sensible objects. And thence arise certain prejudices, for the removing of which, it will be convenient to distinguish them into absolute and relative, true and apparent, mathematical and common.

I. Absolute, true, and mathematical time, of itself, and from its own nature flows equably without regard to anything external, and by another name is called duration: relative, apparent, and common time, is some sensible and external (whether accurate or unequable) measure of duration by the means of motion, which is commonly used instead of true time; such as an hour, a day, a month, a year.

II. Absolute space, in its own nature, without regard to anything external, remains always similar and immovable. Relative space is some movable dimension or measure of the absolute spaces; which our senses determine by its position to bodies; and which is vulgarly taken for immovable space; such is the dimension of a subterraneous, an aereal, or celestial space, determined by its position in respect of the earth. Absolute and relative space, are the same in figure and magnitude; but they do not remain always numerically the same. For if the earth, for instance, moves, a space of our air, which relatively and in respect of the earth remains always the same, will at one time be one part of the absolute space into which the air passes; at another time it will be another part of the same, and so absolutely understood, it will be perpetually mutable.

III. Place is a part of space which a body takes up, and is according to the space, either absolute or relative. I say, a part of space; not the situation, nor the external surface of the body. For the places of equal solids are always equal; but their superfices, by reason of their dissimilar figures, are often unequal. Positions properly have no quantity, nor are they so much the places themselves, as the properties of places. The motion of the whole is the same thing with the sum of the motions of the parts; that is, the translation of the whole, out of its place, is the same thing with the sum of the translations of the parts out of their places; and therefore the place of the whole is the same thing with the sum of the places of the parts, and for that reason, it is internal, and in the whole body.

IV. Absolute motion is the translation of a body from one absolute place into another; and relative motion, the translation from one relative place into another. Thus in a ship under sail, the relative place of a body is that part of the ship which the body possesses; or that part of its cavity which the body fills, and which therefore moves together with the ship: and relative rest is the continuance of the body in the same part of the ship, or of its cavity. But real, absolute rest, is the continuance of the body in the same part of that immovable space, in which the ship itself, its cavity, and all that it contains, is moved. Wherefore, if the earth is really at rest, the body, which relatively rests in the ship, will really and absolutely move with the same velocity which the ship has on the earth. But if the earth also moves, the true and absolute motion of the body will arise, partly from the true motion of the earth, in immovable space; partly from the relative motion of the ship on the earth; and if the body moves also relatively in the ship; its true motion will arise, partly from the true motion of the earth, in immovable space, and partly from the relative motions as well of the ship on the earth, as of the body in the ship; and from these relative motions will arise the relative motion of the body on the earth. As if that part of the earth, where the ship is, was truly moved toward the east, with a velocity of 10010 parts; while the ship itself, with a fresh gale, and full sails, is carried towards the west, with a velocity expressed by 10 of those parts; but a sailor walks in the ship towards the east, with 1 part of the said velocity; then the sailor will be moved truly in immovable space towards the east, with a velocity of 10001 parts, and relatively on the earth towards the west, with a velocity of 9 of those parts.

Absolute time, in astronomy, is distinguished from relative, by the equation or correction of the vulgar time. For the natural days are truly unequal, though they are commonly considered as equal, and used for a measure of time; astronomers correct this inequality for their more accurate deducing of the celestial motions. It may be, that there is no such thing as an equable motion, whereby time may be accurately measured. All motions may be accelerated and retarded; but the true, or equable, progress of absolute time is liable to no change. The duration or perseverance of the existence of things remains the same, whether the motions are swift or slow, or none at all: and therefore it ought to be distinguished from what are only sensible measures thereof; and out of which we collect it, by means of the astronomical equation. The necessity of which equation, for determining the times of a phaenomenon, is evinced as well from the experiments of the pendulum clock, as by eclipses of the satellites of *Jupiter*.

As the order of the parts of time is immutable, so also is the order of the parts of space. Suppose those parts to be moved out of their places, and they will be moved (if the expression may be allowed) out of themselves. For times and spaces are, as it were, the places as well of themselves as of all other things. All things are placed in time as to order of succession; and in space as to order of situation. It is from their essence or nature that they are placed; and that the primary places of things should be moveable, is absurd. These are therefore the absolute places; and translations out of those places, are the only absolute motions.

But because the parts of space cannot be seen, or distinguished from one another by our senses, therefore in their stead we use sensible measures of them. For from the positions and distances of things from any body considered as immovable, we define all places; and then with respect to such places, we estimate all motions, considering bodies as transferred from some of those places into others. And so, instead of absolute places and motions, we use relative ones; and that without any inconvenience in common affairs; but in philosophical disquisitions, we ought to abstract from our senses, and consider things themselves, distinct from what are only sensible measures of them. For it may be that there is no body really at rest, to which the places and motions of others may be referred.

But we may distinguish rest and motion, absolute and relative, one from the other by their properties, causes and effects. It is a property of rest, that bodies really at rest do rest in respect to one another. And therefore as it is possible, that in the remote regions of the fixed stars, or perhaps far beyond them, there may be some body absolutely at rest; but impossible to know, from the position of bodies to one another in our regions whether any of these do keep the same position to that remote body; it follows that absolute rest cannot be determined from the position of bodies in our regions.

It is a property of motion, that the parts, which retain given positions to their wholes, do partake of the motions of those wholes. For all the parts of revolving bodies endeavour to recede from the axis of motion; and the impetus of bodies moving forward, arises from the joint impetus of all the parts. Therefore, if surrounding bodies are moved, those that are relatively at rest within them, will partake of their motion. Upon which account, the true and absolute motion of a body cannot be Jeter- mined by the translation of it from those which only seem to rest; for the external bodies ought not only to appear at rest, but to be really at rest. For otherwise, all included bodies, beside their translation from near the surrounding ones, partake likewise of their true motions ; and though that translation were not made they would not be really at rest, but only seem to be so. For the surrounding bodies stand in the like relation to the surrounded as the exterior part of a whole does to the interior, or as the shell does to the kernel ; but, if the shell moves, the kernel will also move, as being part of the whole, without any removal from near the shell.

A property, near akin to the preceding, is this, that if a place is moved, whatever is placed therein moves along with it ; and therefore a body, which is moved from a place in motion, partakes also of the motion of its place. Upon which account, all motions, from places in motion, are no other than parts of entire and absolute motions; and every entire motion is composed of the motion of the body out of its first place, and the motion of this place out of its place; and so on, until we come to some immovable place, as in the before-mentioned example of the sailor. Wherefore, entire and absolute motions can be no otherwise determined than by immovable places: and for that reason I did before refer those absolute motions to immovable places, but relative ones to movable places. Now no other places are immovable but those that, from infinity to infinity, do all retain the same given position one to another; and upon this account must ever remain unmoved; and do thereby constitute immovable space.

The causes by which true and relative motions are distinguished, one from the other, are the forces impressed upon bodies to generate motion. True motion is neither generated nor altered, but by some force impressed upon the body moved: but relative motion may be generated or altered without any force impressed upon the body. For it is sufficient only to impress some force on other bodies with which the former is compared, that by their giving way, that relation may be changed, in which the relative rest or motion of this other body did consist. Again, true motion suffers always some change from any force impressed upon the moving body; but relative motion docs not necessarily undergo any change by such forces. For if the same forces are likewise impressed on those other bodies, with which the comparison is made, that the relative position may be pre served, then that condition will be preserved in which the relative motion consists. And therefore any relative motion may be changed when the true motion remains unaltered, and the relative may be preserved when the true suffers some change. Upon which accounts; true motion does by no means consist in such relations.

The effects which distinguish absolute from relative motion arc, the forces of receding from the axis of circular motion. For there are no such forces in a circular motion purely relative, but in a true and absolute circular motion, they are greater or less, according to the quantity of the motion. If a vessel, hung: by a long cord, is so often turned about that the cord is strongly twisted, then filled with water, and held at rest together with the water; after, by the sudden action of another force, it is whirled about the contrary way, and while the cord is untwisting itself, the vessel continues for some time in this motion; the surface of the water will at first be plain, as before the vessel began to move: but the vessel; by gradually communicating its motion to the water, will make it begin sensibly to revolve, and recede by little and little from the middle, and ascend to the sides of the vessel, forming itself into a concave figure (as I have experienced), and the swifter the motion becomes, the higher will the water rise, till at last, performing its revolutions in the same times with the vessel, it becomes relatively at rest in it. This ascent of the water shows its endeavour to recede from the axis of its motion; and the true and absolute circular motion of the water, which is here directly contrary to the relative, discovers itself, and may be measured by this endeavour. At first, when the relative motion of the water in the vessel was greatest, it produced no endeavour to recede from the axis; the water showed no tendency to the circumference, nor any ascent towards the sides of the vessel, but remained of a plain surface, and therefore its true circular motion had not yet begun. But afterwards, when the relative motion of the water had decreased, the ascent thereof towards the sides of the vessel proved its endeavour to recede from the axis; and this endeavour showed the real circular motion of the water perpetually increasing, till it had acquired its greatest quantity, when the water rested relatively in the vessel. And therefore this endeavour does not depend upon any translation of the water in respect of the ambient bodies, nor can true circular motion be defined by such translation. There is only one real circular motion of any one revolving body, corresponding to only one power of endeavouring to recede from its axis of motion, as its proper and adequate effect; but relative motions, in one and the same body, are innumerable, according to the various relations it bears to external bodies, and like other relations, are altogether destitute of any real effect, any otherwise than they may perhaps partake of that one only true motion. And therefore in their system who suppose that our heavens, revolving below the sphere of the fixed stars, carry the planets along with them; the several parts of those heavens, and the planets, which are indeed relatively at rest in their heavens, do yet really move. For they change their position one to another (which never happens to bodies truly at rest), and being carried together with their heavens, partake of their motions, and as parts of revolving wholes, endeavour to recede from the axis of their motions.

Wherefore relative quantities are not the quantities themselves, whose names they bear, but those sensible measures of them (either accurate or inaccurate), which are commonly used instead of the measured quantities themselves. And if the meaning of words is to be determined by their use, then by the names time, space, place and motion, their measures are properly to be understood; and the expression will be unusual, and purely mathematical, if the measured quantities themselves are meant. Upon which account, they do strain the sacred writings, who there interpret those words for the measured quantities. Nor do those less defile the purity of mathematical and philosophical truths, who confound real quantities themselves with their relations and vulgar measures.

It is indeed a matter of great difficulty to discover, and effectually to distinguish, the true motions of particular bodies from the apparent; because the parts of that immovable space, in which those motions are performed, do by no means come under the observation of our senses. Yet the thing is not altogether desperate: for we have some arguments to guide us, partly from the apparent motions, which are the differences of the true motions; partly from the forces, which are the causes and effects of the true motions. For instance, if two globes, kept at a given distance one from the other by means of a cord that connects them, were revolved about their common centre of gravity, we might, from the tension of the cord, discover the endeavour of the globes to recede from the axis of their motion, and from thence we might compute the quantity of their circular motions. And then if any equal forces should be impressed at once on the alternate faces of the globes to augment or diminish their circular motions, from the increase or decrease of the tension of the cord, we might infer the increment or decrement of their motions: and thence would be found on what faces those forces ought to be impressed, that the motions of the globes might be most augmented; that is, we might discover their hindermost faces, or those which, in the circular motion, do follow. But the faces which follow being known, and consequently the opposite ones that precede, we should likewise know the determination of their motions. And thus we might find both the quantity and the determination of this circular motion, even in an immense vacuum, where there was nothing external or sensible with which the globes could be compared. But now, if in that space some remote bodies were placed that kept always a given position one to another, as the fixed stars do in our regions, we could not indeed determine from the relative translation of the globes among those bodies, whether the motion did belong to the globes or to the bodies. But if we observed the cord, and found that its tension was that very tension which the motions of the globes required, we might conclude the motion to be in the globes, and the bodies to be at rest; and then, lastly, from the translation of the globes among the bodies, we should find the determination of their motions. But how we are to collect the true motions from their causes, effects, and apparent differences; and, vice versa, how from the motions, either true or apparent, we may come to the knowledge of their causes and effects, shall be explained more at large in the following tract. For to this end it was that I composed it.

Reading 7
Newton: Principia: Axioms

Translation by Andrew Motte

AXIOMS, OR LAWS OF MOTION.

LAW I.

Every body perseveres in its state of rest, or of uniform motion in a right line, unless it is compelled to change that state by forces impressed thereon.

PROJECTILES persevere in their motions, so far as they are not retarded by the resistance of the air, or impelled downwards by the force of gravity. A top, whose parts by their cohesion are perpetually drawn aside from rectilinear motions, does not cease its rotation, otherwise than as it is retarded by the air. The greater bodies of the planets and comets, meeting with less resistance in more free spaces, preserve their motions both progressive and circular for a much longer time.

LAW II.

The alteration of motion is ever proportional to the motive force impressed; and is made in the direction of the right line in which that force is impressed.

If any force generates a motion, a double force will generate double the motion, a triple force triple the motion, whether that force be impressed altogether and at once, or gradually and successively. And this motion (being always directed the same way with the generating force), if the body moved before, is added to or subducted from the former motion, according as they directly conspire with or are directly contrary to each other ; or obliquely joined, when they are oblique, so as to produce a new motion compounded from the determination of both.

LAW III.

To every action there is always opposed an equal reaction: or the mutual actions of two bodies upon each other are always equal, and directed to contrary parts.

Whatever draws or presses another is as much drawn or pressed by that other. If you press a stone with your finger, the finger is also pressed by the stone. If a horse draws a stone tied to a rope, the horse (if I may so say) will be equally drawn back towards the stone: for the distended rope, by the same endeavour to relax or unbend itself, will draw the horse as much towards the stone, as it does the stone towards the horse, and will obstruct the progress of the one as much as it advances that of the other. If a body impinge upon another, and by its force change the motion of the other, that body also (because of the equality of the mutual pressure) will undergo an equal change, in its own motion, towards the contrary part. The changes made by these actions are equal, not in the velocities but in the motions of bodies; that is to say, if the bodies are not hindered by any other impediments. For, because the motions are equally changed, the changes of the velocities made towards contrary parts are reciprocally proportional to the bodies. This law takes place also in attractions, as will be proved in the next scholium.

COROLLARY I.

A body by two forces conjoined will describe the diagonal of a parallelogram, in the same time that it would describe the sides, by those forces apart.

If a body in a given time, by the force M impressed apart in the place A, should with an uniform motion be carried from A to B; and by the force N impressed apart in the same place, should be carried from A to C; complete the parallelogram ABCD, and, by both forces acting together, it will in the same time be carried in the diagonal from A to D. For since the force N acts in the direction of the line AC, parallel to BD, this force (by the second law) will not at all alter the velocity generated by the other force M, by which the body is carried towards the line BD. The body therefore will arrive at the line BD in the same time, whether the force N be impressed or not; and therefore at the end of that time it will be found somewhere in the line BD. By the same argument, at the end of the same time it will be found somewhere in the line CD. Therefore it will be found in the point D, where both lines meet. But it will move in a right line from A to D, by Law I.

COROLLARY III.

The quantity of motion, which is collected by taking the sum of the motions directed towards the same parts, and the difference of those that are directed to contrary parts, suffers no change from the action of bodies among themselves.

For action and its opposite re-action are equal, by Law III, and therefore, by Law II, they produce in the motions equal changes towards opposite parts. Therefore if the motions are directed towards the same parts whatever is added to the motion of the preceding body will be subducted from the motion of that which follows; so that the sum will be the same as before. If the bodies meet, with contrary motions, there will be an equal deduction from the motions of both; and therefore the difference of the motions directed towards opposite parts will remain the same.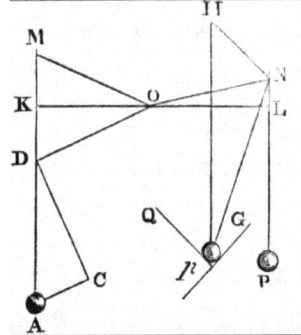

Thus if a spherical body A with two parts of velocity is triple of a spherical body B which follows in the same right line with ten parts of velocity, the motion of A will be to that of B as 6 to 10. Suppose, then, their motions to be of 6 parts and of 10 parts, and the sum will be 16 parts. Therefore, upon the meeting of the bodies, if A acquire 3, 4, or 5 parts of motion, B will lose as man; and therefore after reflexion A will proceed with 9, 10, or 11 parts, and B with 7, 6, or 5 parts;

the sum remaining always of 16 parts as before. If the body A acquire 9, 10, 11, or 12 parts of motion, and therefore after meeting proceed with 15, 16, 17, or 18 parts, the body B, losing so many parts as A has got, will either proceed with 1 part, having lost 9, or stop and remain at rest, as having lost its whole progressive motion of 10 parts ; or it will go back with 1 part, having not only lost its whole motion, but (if 1 may so say) one part more; or it will go back with 2 parts, because a progressive motion of 12 parts is taken off. And so the sums of the conspiring motions 15+1, or 16+0, and the differences of the contrary motions 17-1 and 18-2, will always be equal to 16 parts, as they were before the meeting and reflexion of the bodies. But, the motions being known with which the bodies proceed after reflexion, the velocity of either will be also known, by taking the velocity after to the velocity before reflexion, as the motion after is to the motion before. As in the last case, where the motion of the body A was 6 parts before reflexion and of 18 parts after, and the velocity was of 2 parts before reflexion, the velocity thereof after reflexion will be found to be of 6 parts; by saying, as the parts of motion before to 18 parts after, so are 2 parts of velocity before reflexion to 6 parts after.

But if the bodies are either not spherical, or, moving in different right lines, impinge obliquely one upon the other, and their motions after reflexion are required, in those cases we are first to determine the position of the plane that touches the concurring bodies in the point of concourse, then the motion of each body (by Corol. II) is to be resolved into two, one perpendicular to that plane, and the other parallel to it. This done, because the bodies act upon each other in the direction of a line perpendicular to this plane, the parallel motions are to be retained the same after reflexion as before; and to the perpendicular motions we are to assign equal changes towards the contrary parts; in such manner that the sum of the conspiring and the difference of the contrary motions may remain the same as before. From such kind of reflexions also sometimes arise the circular motions of bodies about their own centres. But these are cases which I do not consider in what follows; and it would be too tedious to demonstrate every particular that relates to this subject.

COROLLARY IV.

The common centre of gravity of two or more bodies does not alter its state of motion or rest by the actions of the bodies among themselves; and therefore the common centre of gravity of all bodies acting upon each other (excluding outward actions and impediments) is either at rest, or moves uniformly in a right line.

For if two points proceed with an uniform motion in right lines, and their distance be divided in a given ratio, the dividing point will be either at rest, or proceed uniformly in a right line. This is demonstrated here after in Lem. XXIII and its Corol., when the points are moved in the same plane; and by a like way of arguing, it may be demonstrated when the points are not moved in the same plane. Therefore if any number of bodies move uniformly in right lines, the common centre of gravity of any two of them is either at rest, or proceeds uniformly in a right line; because the line which connects the centres of those two bodies so moving is divided at that common centre in a given ratio. In like manner the common centre of those two and that of a third body will be either at rest or moving uniformly in aright line because at that centre the distance between the common centre of the two bodies, and the centre of this last, is divided in a given ratio. In like manner the common centre of these three, and of a fourth body, is either at rest, or moves uniformly in a right line; because the distance between the common centre of the three bodies, and the centre of the fourth is there also divided in a given ratio, and so on *ad infinitum.* Therefore, in a system of bodies where there is neither any mutual action among themselves, nor any foreign force impressed upon them from without, and which consequently move uniformly in right lines, the common centre of gravity of them all is either at rest or moves uniformly forward in a right line.

Moreover, in a system of two bodies mutually acting upon each other, since the distances between their centres and the common centre of gravity of both are reciprocally as the bodies, the relative motions of those bodies, whether of approaching to or of receding from that centre, will be equal among themselves. Therefore since the changes which happen to motions are equal and directed to contrary parts, the common centre of those bodies, by their mutual action between themselves, is neither promoted nor retarded, nor suffers any change as to its state of motion or rest. But in a system of several bodies, because the common centre of gravity of any two acting mutually upon each other suffers no change in its state by that action: and much less the common centre of gravity of the others with which that action does not intervene; but the distance between those two centres is divided by the common centre of gravity of all the bodies into parts reciprocally proportional to the total sums of those bodies whose centres they are : and therefore while those two centres retain their state of motion or rest, the common centre of all does also retain its state : it is manifest that the common centre of all never suffers any change in the state of its motion or rest from the actions of any two bodies between themselves. But in such a system all the actions of the bodies among themselves either happen between two bodies, or are composed of actions interchanged between some two bodies; and therefore they do never produce any alteration in the common centre of all as to its state of motion or rest. Wherefore since that centre, when the bodies do not act mutually one upon another, either is at rest or moves uniformly forward in some right line, it will, notwithstanding the mutual actions of the bodies among themselves, always persevere in its state, either of rest, or of proceeding uniformly in a right line, unless it is forced out of this state by the action of some power impressed from without upon the whole system. And therefore the same law takes place in a system consisting of many bodies as in one single body, with regard to their persevering in their state of motion or of rest. For the progressive motion, whether of one single body, or of a whole system of bodies us always to be estimated from the motion of the centre of gravity.

SCHOLIUM.

Hitherto I have laid down such principles as have been received by mathematicians, and are confirmed by abundance of experiments. By the first two Laws and the first two Corollaries, Galileo discovered that the descent of bodies observed the duplicate ratio of the time, and that the motion of projectiles was in the curve of a parabola; experience agreeing with both, unless so far as these motions are a little retarded by the resistance of the air. When a body is falling, the uniform force of its gravity acting equally, impresses, in equal particles of time, equal forces upon that body, and therefore generates equal velocities; and in the whole time impresses a whole force, and generates a whole velocity proportional to the time. And the spaces described in proportional times are as the velocities and the times conjunctly; that is, in a duplicate ratio of the times. And when a body is thrown upwards, its uniform gravity impresses forces and takes off velocities proportional to the times; and the times of ascending to the greatest heights are as the velocities to be taken off, and those heights are as the velocities and the times conjunctly, or in the duplicate ratio of the velocities. And if a body be projected in any direction, the motion arising from its projection is compounded with the motion arising from its gravity. As if the body A by its motion of projection alone could describe in a given time the right line AB, and with its motion of falling alone could describe in the same time the altitude AC; complete the parallelogram ABDC, and the body by that compounded motion will at the end of the time be found in the place D; and the curve line AED, which that body describes, will be a parabola, to which the right line AB will be a tangent in A; and whose ordinate BD will be as the square of the line AB. On the same Laws and Corollaries depend those things which have been demonstrated concerning the times of the vibration of pendulums, and are confirmed by the daily experiments of pendulum clocks. By the same, together with the third Law, Sir Christ. Wren, Dr. Wallis, and Mr. Huygens, the greatest geometers of our times, did severally determine the rules of the congress and reflexion of hard bodies, and much about the same time communicated their discoveries to the Royal Society, exactly agreeing among themselves as to those rules. Dr. Wallis, indeed, was something more early in the publication; then followed Sir Christopher Wren, and, lastly, Mr. Huygens. But Sir Christopher Wren confirmed the truth of the thing before the Royal Society by the experiment of pendulums, which Mr. Mariotte soon after thought fit to explain in a treatise entirely upon that subject. But to bring this experiment to an accurate agreement with the theory, we are to have a due regard as well to the resistance of the air as to the elastic force of the concurring bodies. Let the spherical bodies A, B be suspended by the parallel and equal strings AC, BD, from the centres C, D. About these centres, with those intervals, describe the semicircles EAF, GBH, bisected by the radii CA, DB. Bring the body A to any point R of the arc EAF, and (withdrawing the body B) let it go from thence, and after one oscillation suppose it to return to the point V: then 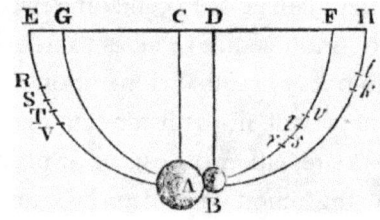 RV will be the retardation arising from the resistance of the air. Of this RV let ST be a fourth part, situated in the middle, to wit, so as RS and TV may be equal, and RS may be to ST as 3 to 2 then will ST represent very nearly the retardation during the descent from S to A. Restore the body B to its place: and, supposing the body A to be let fall from the point S, the velocity thereof in the place of reflexion A, without sensible error, will be the same as if it had descended *in vacuo* from the point T. Upon which account this velocity may be represented by the chord of the arc TA. For it is a proposition well known to geometers, that the velocity of a pendulous body in the lowest point is as the chord of the arc which it has described in its descent. After reflexion, suppose the body A comes to the place s, and the body B to the place k. Withdraw the body B, and find the place v, from which if the body A, being let go, should after one oscillation return to the place r, st may be a fourth part of rv, so placed in the middle thereof as to leave is equal to tv, and let the chord of the arc tA represent the velocity which the body A had in the place A immediately after reflexion. For t will be the true and correct place to which the body A should have ascended, if the resistance of the air had been taken off. In the same way we are to correct the place k to which the body B ascends, by finding the place l to which it should have ascended *in vacuo*. And thus everything may be subjected to experiment, in the same manner as if we were really placed *in vacuo*. These things being done, we are to take the product (if I may so say) of the body A, by the chord of the arc TA (which represents its velocity), that we may have its motion in the place A immediately before reflexion ; and then by the chord of the arc tA, that we may have its motion in the place A immediately after reflexion. And so we are to take the product of the body B by the chord of the arc Bl, that we may have the motion of the same immediately after reflexion. And in like manner, when two bodies are let go together from different places, we are to find the motion of each, as well before as after reflexion; and then we may compare the motions between themselves, and collect the effects of the reflexion. Thus trying the thing with pendulums of ten feet, in unequal as well as equal bodies, and making the bodies to concur after a descent through large spaces, as of 8, 12, or 16 feet, I found always, without an error of 3 inches, that when the bodies concurred together directly, equal changes towards the contrary parts were produced in their motions, and, of consequence, that the action and reaction were always equal. As if the body A impinged upon the body B at rest with 9 parts of motion, and losing 7, proceeded after reflexion with 2, the body B was carried backwards with those 7 parts. If the bodies concurred with contrary motions, A with twelve parts of motion, and B with six, then if A receded with 2, B receded with 8; to wit, with a deduction of 14 parts of motion on each side. For from the motion of A subducting twelve parts, nothing will remain; but subducting 2 parts more, a motion will be generated of 2 parts towards the contrary way; and so, from the motion of the body B of 6 parts, subducting 14 parts, a motion is generated of 8 parts towards the contrary way. But if the bodies were made both to move towards the same way, A, the swifter, with 14 parts of motion, B, the slower, with 5, and after reflexion A went on with 5, B likewise went on with 14 parts; 9 parts being transferred from A to B. And so in other cases. By the congress and collision of bodies, the quantity of motion, collected from the sum of the motions directed towards the same way, or from the difference, of those that were directed towards contrary ways, was never changed. For the error of an inch or two in measures may be easily ascribed to the difficulty of executing everything with accuracy. It was not easy to let go the two pendulums so exactly together that the bodies should impinge one upon the other in the lowermost place AB; nor to mark the places s, and k, to which the bodies ascended after congress. Nay, and some errors, too, might have happened from the unequal density of the parts of the pendulous bodies themselves, and from the irregularity of the texture proceeding from other causes.

But to prevent an objection that may perhaps be alledged against the rule, for the proof of which this experiment was made, as if this rule did suppose that the bodies were either absolutely hard, or at least perfectly elastic (whereas no such bodies are to be found in nature), I must add. that the experiments we have been describing, by no means depending upon that quality of hardness, do succeed as well in soft as in hard bodies. For if the rule is to be tried in bodies not perfectly hard, we are only to diminish the reflexion in such a certain proportion as the quantity of the elastic force requires. By the theory of Wren and Huygens, bodies absolutely hard return one from another with the same velocity with which they meet. But this may be affirmed with more certainty of bodies perfectly elastic. In bodies imperfectly elastic the velocity of the return is to be diminished together with the elastic force; because that force (except when the parts of bodies are bruised by their congress, or suffer some such extension as happens under the strokes of a hammer) is (as far as I can perceive) certain and determined, and makes the bodies to return one from the other with a relative velocity, which is in a given ratio to that relative velocity with which they met. This I tried in balls of wool, made up tightly, and strongly compressed. For, first, by letting go the pendulous bodies, and measuring their reflexion, I determined the quantity of their elastic force; and then, according to this force, estimated the reflexions that ought to happen in other cases of congress.

And with this computation other experiments made afterwards did accordingly agree; the balls always receding one from the other with a relative velocity, which was to the relative velocity with which they met as about 5 to 9. Balls of steel returned with almost the same velocity: those of cork with a velocity something less; but in balls of glass the proportion was as about 15 to 16. And thus the third Law, so far as it regards percussions and reflexions, is proved by a theory exactly agreeing with experience.

In attractions, I briefly demonstrate the thing after this manner. Suppose an obstacle is interposed to hinder the congress of any two bodies A, B, mutually attracting one the other: then if either body, as A, is more attracted towards the other body B, than that other body B is towards the first body A, the obstacle will be more strongly urged by the pressure of the body A than by the pressure of the body B, and therefore will not remain in equilibrio: but the stronger pressure will prevail, and will make the system of the two bodies, together with the obstacle, to move directly towards the parts on which B lies ; arid in free spaces, to go forward *in infinitum* with a motion perpetually accelerated ; which is absurd and contrary to the first Law. For, by the first Law, the system ought to persevere in its state of rest, or of moving uniformly forward in a right line: and therefore the bodies must equally press the obstacle, and be equally attracted one by the other. I made the experiment on the loadstone and iron. If these, placed apart in proper vessels, are made to float by one another in standing water, neither of them will propel the other; but, by being equally attracted, they will sustain each other's pressure, and rest at last in an equilibrium.

So the gravitation betwixt the earth and its parts is mutual. Let the earth FI be cut by any plane EG into two parts EGF and EGI, and their weights one towards the other will be mutually equal. For if by another plane HK, parallel to the former EG, the greater part EGI is cut into two parts EGKH and HKI, whereof HKI is equal to the part EFG, first cut off, it is evident that the middle part EGKH, will have no propension by its proper weight towards either side, but will hang as it were, and rest in an equilibrium betwixt both. 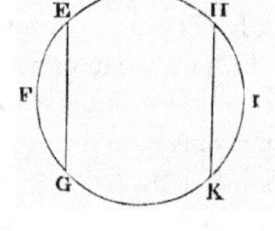 But the one extreme part HKI will with its whole weight bear upon and press the middle part towards the other extreme part EGF: and therefore the force with which EGI, the sum of the parts HKI and EGKH, tends towards the third part EGF, is equal to the weight of the part HKI, that is, to the weight of the third part EGF. And therefore the weights of the two parts EGI and EGF, one towards the other, are equal, as I was to prove. And indeed if those weights were not equal, the whole earth floating in the non- resisting aether would give way to the greater weight, and, retiring from it, would be carried off in infinitum.

And as those bodies are equipollent in the congress and reflexion, whose velocities are reciprocally as their innate forces, so in the use of mechanic instruments those agents are equipollent, and mutually sustain each the contrary pressure of the other, whose velocities, estimated according to the determination of the forces, are reciprocally as the forces.

So those weights are of equal force to move the arms of a balance; which during the play of the balance are reciprocally as their velocities upwards and downwards; that is, if the ascent or descent is direct, those weights are of equal force, which are reciprocally as the distances of the points at which they are suspended from the axis of the balance: but if they are turned aside by the interposition of oblique planes, or other obstacles, and made to ascend or descend obliquely, those bodies will be equipollent, which are reciprocally as the heights of their ascent and descent taken according to the perpendicular; and that on account of the determination of gravity downwards.

And in like manner in the pully, or in a combination of pullies, the force of a hand drawing the rope directly, which is to the weight, whether ascending directly or obliquely, as the velocity of the perpendicular ascent of the weight to the velocity of the hand that draws the rope, will sustain the weight.

In clocks and such like instruments, made up from a combination of wheels, the contrary forces that promote and impede the motion of the wheels, if they are reciprocally as the velocities of the parts of the wheel on which they are impressed, will mutually sustain the one the other.

The force of the screw to press a body is to the force of the hand that turns the handles by which it is moved as the circular velocity of the handle in that part where it is impelled by the hand is to the progressive velocity of the screw towards the pressed body.

The forces by which the wedge presses or drives the two parts of the wood it cleaves are to the force of the mallet upon the wedge as the progress of the wedge in the direction of the force impressed upon it by the mallet is to the velocity with which the parts of the wood yield to the wedge, in the direction of lines perpendicular to the sides of the wedge. And the like account is to be given of all machines.

The power and use of machines consist only in this, that by diminishing the velocity we may augment the force, and the contrary: from whence in all sorts of proper machines, we have the solution of this problem; to move a given weight with a given power, or with a given force to overcome any other given resistance. For if machines are so contrived that the velocities of the agent and resistant are reciprocally as their forces, the agent will just sustain the resistant, but with a greater disparity of velocity will overcome it. So that if the disparity of velocities is so great as to overcome all that resistance which commonly arises either from the attrition of contiguous bodies as they slide by one another, or from the cohesion of continuous bodies that are to be separated, or from the weights of bodies to be raised, the excess of the force remaining, after all those resistances are overcome, will produce an acceleration of motion proportional thereto, as well in the parts of {he machine as in the resisting body. But to treat of mechanics is not my present business. I was only willing to show by those examples the great extent and certainty of the third Law of motion. For if we estimate the action of the agent from its force and velocity conjunctly, and likewise the reaction of the impediment conjunctly from the velocities of its several parts, and from the forces of resistance arising from the attrition, cohesion, weight, and acceleration of those parts, the action and reaction in the use of all sorts of machines will be found always equal to one another. And so far as the action is propagated by the intervening instruments, and at last impressed upon the resisting body, the ultimate determination of the action will be always contrary to the determination of the reaction.

Reading 8
Maxwell: Theory of Heat

Chapter IV: On Work and Energy

2320   Work is done when resistance is overcome, and the quantity of work done is measured by the product of the resisting force and the distance through which that force is overcome.

Thus, if a body whose mass is one pound is lifted one foot high in opposition to the force of gravity, a certain amount of work is done, and this quantity is known among engineers as a foot-pound.

2325   If a body whose mass is twenty pounds is lifted ten feet, this might be done by taking one of the pounds and raising it first one foot and then another till it had risen ten feet, and then doing the same with each of the remaining pounds, so that the quantity of work called a foot-pound is performed 200 times in raising twenty pounds ten feet. Hence the work done in lifting a body is found by multiplying the weight of the body in pounds by the height in feet. The result is the work
2330   in foot-pounds.

The foot-pound is a *gravitation* measure, depending on ·the force of gravity at the place. To reduce it to absolute measure we must multiply the number of foot-pounds by the force of gravity at the place.

The work done when we raise a heavy body is done in overcoming the attraction of the earth.
2335   Work is also done when we draw asunder two magnets which attract each other, when we draw out an elastic cord, when we compress air, and, in general, when we apply force to anything which moves in the direction of the force.

There is one case of the application of force to a moving body which is of great importance, namely, when the force is employed in changing the velocity of the body.

2340   Suppose a body whose mass is M (M pounds or M grammes) to be moving in, a certain direction with a velocity which we shall call v, and let a force. which we shall call F, be applied to the body in the direction of its motion. Let us consider the effect of this force acting on the body for a very small time T, during which the body moves through the space s, and at the end of which its velocity is v'.

2345   To ascertain the magnitude of the force F, let us consider the momentum which it produces in the body, and the time during which the momentum is produced.

The momentum of the beginning of the time T was Mv, and at the end of the time T it was Mv', so that the momentum produced by the force F acting for the time T is Mv' - Mv.

But since forces are measured by the momentum produced in unit of time, the momentum
2350   produced by F in one unit of time is F, and the momentum produced by F in T units of time is FT. Since the two values are equal,

$$FT = M(v' - v).$$

This is one form of the fundamental equation of dynamics. If we define the impulse of a force as the average value of the force multiplied by the time during which it acts, then this equation may be expressed in words by saying that the impulse of a force is equal to the momentum produced by it.

We have next to find s, the space described by the body during the time T. If the velocity had been uniform, the space described would have been the product of the time by the velocity. When the velocity is not uniform the time must be multiplied by the mean or average velocity to get the space described. In both these cases in which average force or average velocity is mentioned, the time is supposed to be subdivided into a number of equal parts, and the average is taken of the force or of the velocity for all these divisions of the time. In the present case, in which the time considered is so small that the change of velocity' is also small, the average velocity during the time T may be taken as the arithmetical mean of the velocities at the beginning and at the end of the time, or $½(v + v')$.

Hence the space described is

$$s = ½(v + v')T.$$

This may be considered as a kinematical equation, since it depends on the nature of motion only, and not on that of the moving body.

If we multiply together these two equations we get

$$FTs = ½M(v'^2 - v^2)T ;$$

and if we divide by T we find

$$Fs = ½Mv'^2 - ½Mv^2.$$

Now Fs is the work done by the force F acting on the body while it moves in the direction of F through a space s. If we also denote $½Mv^2$, the mass of the body multiplied by half the square of its velocity, by the expression the kinetic energy of the body, then $½Mv'^2$ will be the kinetic energy after the action of the force F through a space s.

We may now express the .equation in words by saying that the work done by the force F in setting the body in motion is measured by the increase of kinetic energy during the time that the force acts. We have proved that this is true when the interval of time during which the force acts is so small that we may consider the mean velocity during that time as equal to the arithmetical mean of the velocities at the beginning and end of the time. This assumption, which is exactly true when the force is uniform, is approximately true in every case when the time considered is small enough.

By dividing the whole time of action of the force into small parts, and proving that in each of these the work done by the force is equal to the increase of kinetic energy of the body, we may, by adding the different portions of the work and the different increments of energy, arrive at the result that the total work done by the force is equal to the total increase of kinetic energy.

If the force acts on the body in the direction opposite to the motion, the kinetic energy of the body will be diminished instead of increased, and the force, instead of doing work on the body, will be a resistance which the body in its motion overcomes. Hence a moving body can do work in overcoming resistance as long as it is in motion, and the work done by the moving body is equal to the diminution of its kinetic energy, till, when the body is brought to rest, the whole work it has done is equal to the whole kinetic energy which it had at first.

We now see the appropriateness of the name kinetic energy, which we have hitherto used merely as a name for the product $\frac{1}{2}Mv^2$. For the energy of a body may be defined as the capacity which it has of doing work, and is measured by the quantity of work which it can do. The kinetic energy of a body is the energy which it has in virtue of being in *motion*, and we have just shown that its value may be found by multiplying the mass of the body by half the square of the velocity.

In our investigation we have, for the sake of simplicity, supposed the force to act in the same direction as the motion. To make the proof perfectly general, as it is given in treatises on dynamics, we have only to resolve the actual force into two parts, one in the direction of the motion and the other at right angles to it~ and to observe that the part at right angles to the motion can neither do any work on the body nor change the velocity or the kinetic energy, so that the whole effect, whether of work or of alteration of kinetic energy, depends on the part of the force which is in the direction of the motion.

The student, if not familiar with this subject, should refer to some treatise on dynamics, and compare the investigation there given with the outline of the reasoning given above. Our object at present is to fix in our minds what is meant by Work and Energy.

The great importance of giving a name to the quantity which we call Kinetic Energy seems to have been first recognised by Leibnitz, who gave to the product of the mass by the square of the velocity the name of *Vis Viva*. This is twice the kinetic energy.

Newton, in a scholium to his Third Law or. Motion, has stated the relation between work and kinetic energy in a manner so perfect that it cannot be improved, but at the same time with so little apparent effort or desire to attract attention that no one seems to have been struck with the great importance of the passage till it was pointed out recently by Thomson and Tait.

The use of the term Energy, in a scientific sense, to express the quantity of work a body can do, was introduced by Dr. Young (' Lectures on Natural Philosophy,' Lecture VIII.).

The energy of a system of bodies acting on one another with forces depending on their relative positions is due partly to their motion, and partly to their relative position.

That part which is due to their motion was called Actual Energy by Rankine, and Kinetic Energy by Thomson and Tait. That part which is due to their relative position depends upon the work which the various forces would do if the bodies were to yield to the action of these forces. This is called the Sum of the Tensions by Helmholtz, in his celebrated memoir on the 'Conservation of Force.' Thomson called it Statical Energy, and Rankine introduced the term Potential Energy, a very felicitous name, since it not only signifies the energy which the system has not in possession, but only has the power to acquire, but it also indicates that it is to be found from what is called (on other grounds) the Potential Function.

Thus when a heavy body has been lifted to a certain height above the earth's surface, the system of two bodies, it and the earth, have potential energy equal to the work which would be done if the heavy body were allowed to descend till it is stopped by the surface of the earth.

If the body were allowed to fall freely, it would acquire velocity, and the kinetic energy acquired would be exactly equal to the potential energy lost in the same time.

It is proved in treatises on dynamics that if, in any system of bodies, the force which acts between any two bodies is in the line joining them, and depends only on their distance, and not on the way in which they are moving at the time, then if no other forces act on the system, the sum of the potential and kinetic energy of all the bodies of the system will always remain the same.

This principle is called the Principle of the Conservation of Energy; it is of great importance in all branches of science, and the recent advances in the science of heat have been chiefly due to the application of this principle.

We cannot indeed assume, without evidence of a satisfactory nature, that the mutual action between any two parts of a real body must always be in the line joining them, and must depend only on their distance. We know that this is the case with respect to the attraction of bodies at a distance, but we cannot make any such assumption concerning the internal forces of bodies of whose internal constitution we know next to nothing.

We cannot even assert that all energy must be either potential or kinetic, though we may not be able to conceive any other form. Nevertheless, the principle has been demonstrated by dynamical reasoning to be absolutely true for systems fulfilling certain conditions, and it has been proved by experiment to be true within the limits of error of observation, in cases where the energy takes the forms of heat,. magnetisation, electrification, &c., so that the following statement is one which, if we cannot absolutely affirm its necessary truth, is worthy of being carefully tested, and traced into all the conclusions which are implied in it.

GENERAL STATEMENT OF THE CONSERVATION OF ENERGY.

The total energy of any body or system of bodies is a quantity which can neither be increased nor diminished by any mutual action of these bodies, though it may be transformed into any of the forms of which energy is susceptible.'

If by the application of mechanical force, heat, or any other kind of action to a body, or system of bodies, it is made to pass through any series of changes, and at last to return in all respects to its original state, then the energy communicated to the system during this cycle of operations must be equal to the energy which the system communicates to other bodies during the cycle. For the system is in all respects the same at the beginning and at the end of the cycle, and in particular it has the same amount of energy in it; and therefore, since no internal action of the system can either produce or destroy energy, the quantity of energy which enters the system must be equal to that which leaves it during the cycle.

The reason for believing heat not to be a substance is that it can be generated, so that the quantity of it may be increased to any extent, and it can also be destroyed, though this operation requires certain conditions to be fulfilled. The reason for believing heat to be a form of energy is that heat may be generated by the application of work, and that for every unit of heat which is generated a certain quantity of mechanical energy disappears. Besides, work may be done by the action of heat, and for every foot-pound of work so done a certain quantity of heat is put out of existence.

Now when the appearance of one thing is strictly connected with the disappearance of another, so that the amount which exists of the one thing depends on and can be calculated from the amount of the other which has disappeared, we conclude that the one has been formed at the expense of the other, and that they are both forms of the same thing.

Hence we conclude that heat is energy in a peculiar form. The reasons for believing heat as it exists in a hot body to be in the form of kinetic energy-that is, that the particles of the hot body are in actual though invisible motion-will be discussed afterwards.

Reading 9
Huygens: Treatise on Light

Translated By Silvanus P. Thompson

Preface

I wrote this Treatise during my sojourn in France twelve years ago, and I communicated it in the year 1678 to the learned persons who then composed the Royal Academy of Science, to the membership of which the King had done me the honour of calling, me. There will be seen in it demonstrations of those kinds which do not produce as great a certitude as those of Geometry, and which even differ much therefrom, since whereas the Geometers prove their Propositions by fixed and incontestable Principles, here the Principles are verified by the conclusions to be drawn from them; the nature of these things not allowing of this being done otherwise.

It is always possible to attain thereby to a degree of probability which very often is scarcely less than complete proof. To wit, when things which have been demonstrated by the Principles that have been assumed correspond perfectly to the phenomena which experiment has brought under observation; especially when there are a great number of them, and further, principally, when one can imagine and foresee new phenomena which ought to follow from the hypotheses which one employs, and when one finds that therein the fact corresponds to our prevision. But if all these proofs of probability are met with in that which I propose to discuss, as it seems to me they are, this ought to be a very strong confirmation of the success of my inquiry; and it must be ill if the facts are not pretty much as I represent them. I would believe then that those who love to know the Causes of things and who are able to admire the marvels of Light, will find some satisfaction in these various speculations regarding it, and in the new explanation of its famous property which is the main foundation of the construction of our eyes and of those great inventions which extend so vastly the use of them.

Chapter 1: On Rays Propagated in Straight Lines

As happens in all the sciences in which Geometry is applied to matter, the demonstrations concerning Optics are founded on truths drawn from experience. Such are that the rays of light are propagated in straight lines; that the angles of reflexion and of incidence are equal; and that in refraction the ray is bent according to the law of sines, now so well known, and which is no less certain than the preceding laws.

The majority of those who have written touching the various parts of Optics have contented themselves with presuming these truths. But some, more inquiring, have desired to investigate the origin and the causes, considering these to be in themselves wonderful effects of Nature. In which they advanced some ingenious things, but not however such that the most intelligent folk do not wish for better and more satisfactory explanations. Wherefore I here desire to propound what I have meditated on the subject, so as to contribute as much as I can to the explanation of this department of Natural Science, which, not without reason, is reputed to be one of its most difficult parts. I recognize myself to be much indebted to those who were the first to begin to dissipate the strange obscurity in which these things were enveloped, and to give us hope that they might be explained by intelligible reasoning. But, on the other hand I am astonished also that even here these have often been willing to offer, as assured and demonstrative, reasonings which were far from conclusive. For I do not find that any one has yet given a probable explanation of the first and most notable phenomena of light, namely why it is not propagated except in straight lines, and how visible rays, coming from an infinitude of diverse places, cross one another without hindering one another in any way.

It is inconceivable to doubt that light consists in the motion of some sort of matter. For whether one considers its production, one sees that here upon the Earth it is chiefly engendered by fire and flame which contain without doubt bodies that are in rapid motion, since they dissolve and melt many other bodies, even the most solid; or whether one considers its effects, one sees that when light is collected, as by concave mirrors, it has the property of burning as a fire does, that is to say it disunites the particles of bodies. This is assuredly the mark of motion, at least in the true Philosophy, in which one conceives the causes of all natural effects in terms of mechanical motions. This, in my opinion, we must necessarily do, or else renounce all hopes of ever comprehending anything in Physics.

And as, according to this Philosophy, one holds as certain that the sensation of sight is excited only by the impression of some movement of a kind of matter which acts on the nerves at the back of our eyes, there is here yet one reason more for believing that light consists in a movement of the matter which exists between us and the luminous body.

Further, when one considers the extreme speed with which light spreads on every side, and how, when it comes from different regions, even from those directly opposite, the rays traverse one another without hindrance, one may well understand that when we see a luminous object, it cannot be by any transport of matter coming to us from this object, in the way in which a shot or an arrow traverses the air; for assuredly that would too greatly impugn these two properties of light, especially the second of them. It is then in some other way that light spreads; and that which can lead us to comprehend it is the knowledge which we have of the spreading of Sound in the air.

We know that by means of the air, which is an invisible and impalpable body, Sound spreads around the spot where it has been produced, by a movement which is passed on successively from one part of the air to another; and that the spreading of this movement, taking place equally rapidly on all sides, ought to form spherical surfaces ever enlarging and which strike our ears. Now there is no doubt at all that light also comes from the luminous body to our eyes by some movement impressed on the matter which is between the two; since, as we have already seen, it cannot be by the transport of a body which passes from one to the other. If, in addition, light takes time for its passage—which we are now going to examine—it will follow that this movement, impressed on the intervening matter, is successive; and consequently it spreads, as Sound does, by spherical surfaces and waves: for I call them waves from their resemblance to those which are seen to be formed in water when a stone is thrown into it, and which present a successive spreading as circles, though these arise from another cause, and are only in a flat surface.

As regards the different modes in which I have said the movements of Sound and of Light are communicated, one may sufficiently comprehend how this occurs in the case of Sound if one considers that the air is of such a nature that it can be compressed and reduced to a much smaller space than that which it ordinarily occupies. And in proportion as it is compressed the more does it exert an effort to regain its volume; for this property along with its penetrability, which remains notwithstanding its compression, seems to prove that it is made up of small bodies which float about and which are agitated very rapidly in the ethereal matter composed of much smaller parts. So that the cause of the spreading of Sound is the effort which these little bodies make in collisions with one another, to regain freedom when they are a little more squeezed together in the circuit of these waves than elsewhere.

But the extreme velocity of Light, and other properties which it has, cannot admit of such a propagation of motion, and I am about to show here the way in which I conceive it must occur. For this, it is needful to explain the property which hard bodies must possess to transmit movement from one to another.

When one takes a number of spheres of equal size, made of some very hard substance, and arranges them in a straight line, so that they touch one another, one finds, on striking with a similar sphere against the first of these spheres, that the motion passes as in an instant to the last of them, which separates itself from the row, without one's being able to perceive that the others have been stirred. And even that one which was used to strike remains motionless with them. Whence one sees that the movement passes with an extreme velocity which is the greater, the greater the hardness of the substance of the spheres.

Now in applying this kind of movement to that which produces Light there is nothing to hinder us from estimating the particles of the ether to be of a substance as nearly approaching to perfect hardness and possessing a springiness as prompt as we choose. It is not necessary to examine here the causes of this hardness, or of that springiness, the consideration of which would lead us too far from our subject. I will say, however, in passing that we may conceive that the particles of the ether, notwithstanding their smallness, are in turn composed of other parts and that their springiness consists in the very rapid movement of a subtle matter which penetrates them from every side and constrains their structure to assume such a disposition as to give to this fluid matter the most overt and easy passage possible. This accords with the explanation which Mr. DesCartes gives for the spring, though I do not, like him, suppose the pores to be in the form of round hollow canals. And it must not be thought that in this there is anything absurd or impossible, it being on the contrary quite credible that it is this infinite series of different sizes of corpuscles, having different degrees of velocity, of which Nature makes use to produce so many marvellous effects.

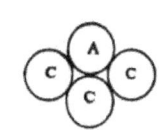

And it must be known that although the particles of the ether are not ranged thus in straight lines, as in our row of spheres, but confusedly, so that one of them touches several others, this does not hinder them from transmitting their movement and from spreading it always forward. As to this it is to be remarked that there is a law of motion serving for this propagation, and verifiable by experiment. It is that when a sphere, such as A here, touches several other similar spheres CCC, if it is struck by another sphere B in such a way as to exert an impulse against all the spheres CCC which touch it, it transmits to them the whole of its movement, and remains after that motionless like the sphere B. And without supposing that the ethereal particles are of spherical form (for I see indeed no need to suppose them so) one may well understand that this property of communicating an impulse does not fail to contribute to the aforesaid propagation of movement.

I have then shown in what manner one may conceive Light to spread successively, by spherical waves, and how it is possible that this spreading is accomplished with as great a velocity as that which experiments and celestial observations demand. Whence it may be further remarked that although the particles are supposed to be in continual movement (for there are many reasons for this) the successive propagation of the waves cannot be hindered by this; because the propagation consists nowise in the transport of those particles but merely in a small agitation which they cannot help communicating to those surrounding, notwithstanding any movement which may act on them causing them to be changing positions amongst themselves.

But what may at first appear full strange and even incredible is that the undulations produced by such small movements and corpuscles, should spread to such immense distances; as for example from the Sun or from the Stars to us. For the force of these waves must grow feeble in proportion as they move away from their origin, so that the action of each one in particular will without doubt become incapable of making itself felt to our sight. But one will cease to be astonished by considering how at a great distance from the luminous body an infinity of waves, though they have issued from different points of this body, unite together in such a way that they sensibly compose one single wave only, which, consequently, ought to have enough force to make itself felt. Thus this infinite number of waves which originate at the same instant from all points of a fixed star, big it may be as the Sun, make practically only one single wave which may well have force enough to produce an impression on our eyes. Moreover from each luminous point there may come many thousands of waves in the smallest imaginable time, by the frequent percussion of the corpuscles which strike the Ether at these points: which further contributes to rendering their action more sensible.

There is the further consideration in the emanation of these waves, that each particle of matter in which a wave spreads, ought not to communicate its motion only to the next particle which is in the straight line drawn from the luminous point, but that it also imparts some of it necessarily to all the others which touch it and which oppose themselves to its movement. So it arises that around each particle there is made a wave of which that particle is the centre. Thus if DCF is a wave emanating from the luminous point A, which is its centre, the particle B, one of those comprised within the sphere DCF, will have made its particular or partial wave KCL, which will touch the wave DCF at C at the same moment that the principal wave emanating from the point A has arrived at DCF; and it is clear that it will be only the region C of the wave KCL which will touch the wave DCF, to wit, that which is in the straight line drawn through AB. Similarly the other particles of the sphere DCF, such as *bb*, *dd*, etc., will each make its own wave. But each of these waves can be infinitely feeble only as compared with the wave DCF, to the composition of which all the others contribute by the part of their surface which is most distant from the centre A.

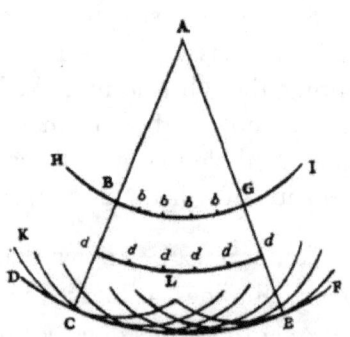

One sees, in addition, that the wave DCF is determined by the distance attained in a certain space of time by the movement which started from the point A; there being no movement beyond this wave, though there will be in the space which it encloses, namely in parts of the particular waves, those parts which do not touch the sphere DCF. And all this ought not to seem fraught with too much minuteness or subtlety, since we shall see in the sequel that all the properties of Light, and everything pertaining to its reflexion and its refraction, can be explained in principle by this means.

2640 This is a matter which has been quite unknown to those who hitherto have begun to consider the waves of light, amongst whom are Mr. Hooke in his *Micrographia*, and Father Pardies, who, in a treatise of which he let me see a portion, and which he was unable to complete as he died shortly afterward, had undertaken to prove by these waves the effects of reflexion and refraction. But the chief foundation, which consists in the remark I have just made, was lacking in his demonstrations;
2645 and for the rest he had opinions very different from mine, as may be will appear some day if his writing has been preserved.

To come to the properties of Light. We remark first that each portion of a wave ought to spread in such a way that its extremities lie always between the same straight lines drawn from the luminous point. Thus the portion BG of the wave, having the luminous point A as its centre, will spread into
2650 the arc CE bounded by the straight lines ABC, AGE. For although the particular waves produced by the particles comprised within the space CAE spread also outside this space, they yet do not concur at the same instant to compose a wave which terminates the movement, as they do precisely at the circumference CE, which is their common tangent.

And hence one sees the reason why light, at least if its rays are not reflected or broken, spreads
2655 only by straight lines, so that it illuminates no object except when the path from its source to that object is open along such lines.

For if, for example, there were an opening BG, limited by opaque bodies BH, GI, the wave of light which issues from the point A will always be terminated by the straight lines AC, AE, as has just been shown; the parts of the partial waves which spread outside the space ACE being too feeble
2660 to produce light there.

Chapter II: On Reflection

Having explained the effects of waves of light which spread in a homogeneous matter, we will examine next that which happens to them on encountering other bodies. We will first make evident how the Reflexion of light is explained by these same waves, and why it preserves equality of angles.

Let there be a surface AB; plane and polished, of some metal, glass, or other body, which at first
2665 I will consider as perfectly uniform (reserving to myself to deal at the end of this demonstration with the inequalities from which it cannot be exempt), and let a line AC, inclined to AD, represent a portion of a wave of light, the centre of which is so distant that this portion AC may be considered as a straight line; for I consider all this as in one plane, imagining to myself that the plane in which this figure is, cuts the sphere of the wave through its centre and intersects the plane AB at right
2670 angles. This explanation will suffice once for all.

The piece C of the wave AC, will in a certain space of time advance as far as the plane AB at B, following the straight line CB, which may be supposed to come from the luminous centre, and which in consequence is perpendicular to AC. Now in this
2675 same space of time the portion A of the same wave, which has been hindered from communicating its movement beyond the plane AB, or at least partly so, ought to have continued its movement in the matter which is above this plane, and this along a distance equal to CB, making its own partial spherical

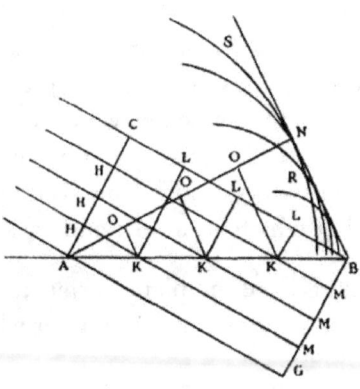

wave, according to what has been said above. Which wave is here represented by the circumference SNR, the centre of which is A, and its semi-diameter AN equal to CB.

If one considers further the other pieces H of the wave AC, it appears that they will not only have reached the surface AB by straight lines HK parallel to CB, but that in addition they will have generated in the transparent air, from the centres K, K, K, particular spherical waves, represented here by circumferences the semi-diameters of which are equal to KM, that is to say to the continuations of HK as far as the line BG parallel to AC. But all these circumferences have as a common tangent the straight line BN, namely the same which is drawn from B as a tangent to the first of the circles, of which A is the centre, and AN the semi-diameter equal to BC, as is easy to see.

It is then the line BN (comprised between B and the point N where the perpendicular from the point A falls) which is as it were formed by all these circumferences, and which terminates the movement which is made by the reflexion of the wave AC; and it is also the place where the movement occurs in much greater quantity than anywhere else. Wherefore, according to that which has been explained, BN is the propagation of the wave AC at the moment when the piece C of it has arrived at B. For there is no other line which like BN is a common tangent to all the aforesaid circles, except BG below the plane AB; which line BG would be the propagation of the wave if the movement could have spread in a medium homogeneous with that which is above the plane. And if one wishes to see how the wave AC has come successively to BN, one has only to draw in the same figure the straight lines KO parallel to BN, and the straight lines KL parallel to AC. Thus one will see that the straight wave AC has become broken up into all the OKL parts successively, and that it has become straight again at NB.

Now it is apparent here that the angle of reflexion is made equal to the angle of incidence. For the triangles ACB, BNA being rectangular and having the side AB common, and the side CB equal to NA, it follows that the angles opposite to these sides will be equal, and therefore also the angles CBA, NAB. But as CB, perpendicular to CA, marks the direction of the incident ray, so AN, perpendicular to the wave BN, marks the direction of the reflected ray; hence these rays are equally inclined to the plane AB.

Chapter III: On Refraction

I will finish this theory of refraction by demonstrating a remarkable proposition which depends on it; namely, that a ray of light in order to go from one point to another, when these points are in different media, is refracted in such wise at the plane surface which joins these two media that it employs the least possible time: and exactly the same happens in the case of reflexion against a plane surface. Mr. Fermat was the first to propound this property of refraction, holding with us, and directly counter to the opinion of Mr. DesCartes, that light passes more slowly through glass and water than through air. But he assumed besides this a constant ratio of Sines, which we have just proved by these different degrees of velocity alone: or rather, what is equivalent, he assumed not only that the velocities were different but that the light took the least time possible for its passage, and thence deduced the constant ratio of the Sines. His demonstration, which may be seen in his printed works, and in the volume of letters of Mr. DesCartes, is very long; wherefore I give here another which is simpler and easier.

Let KF be the plane surface; A the point in the medium which the light traverses more easily, as the air; C the point in the other which is more difficult to penetrate, as water. And suppose that a ray has come from A, by B, to C, having been refracted at B according to the law demonstrated a little before; that is to say that, having drawn PBQ, which cuts the plane at right angles, let the sine of the angle ABP have to the sine of the angle CBQ the same ratio as the velocity of light in the medium where A is to the velocity of light in the medium where C is. It is to be shown that the time of passage of light along AB and BC taken together, is the shortest that can be. Let us assume that it may have come by other lines, and, in the first place, along AF, FC, so that the point of refraction F may be further from B than the point A; and let AO be a line perpendicular to AB, and FO parallel to AB; BH perpendicular to FO, and FG to BC.

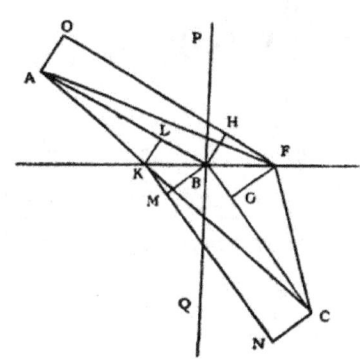

Since then the angle HBF is equal to PBA, and the angle BFG equal to QBC, it follows that the sine of the angle HBF will also have the same ratio to the sine of BFG, as the velocity of light in the medium A is to its velocity in the medium C. But these sines are the straight lines HF, BG, if we take BF as the semi-diameter of a circle. Then these lines HF, BG, will bear to one another the said ratio of the velocities. And, therefore, the time of the light along HF, supposing that the ray had been OF, would be equal to the time along BG in the interior of the medium C. But the time along AB is equal to the time along OH; therefore the time along OF is equal to the time along AB, BG. Again the time along FC is greater than that along GC; then the time along OFC will be longer than that along ABC. But AF is longer than OF, then the time along AFC will by just so much more exceed the time along ABC.

Now let us assume that the ray has come from A to C along AK, KC; the point of refraction K being nearer to A than the point B is; and let CN be the perpendicular upon BC, KN parallel to BC: BM perpendicular upon KN, and KL upon BA.

Here BL and KM are the sines of angles BKL, KBM; that is to say, of the angles PBA, QBC; and therefore they are to one another as the velocity of light in the medium A is to the velocity in the medium C. Then the time along LB is equal to the time along KM; and since the time along BC is equal to the time along MN, the time along LBC will be equal to the time along KMN. But the time along AK is longer than that along AL: hence the time along AKN is longer than that along ABC. And KC being longer than KN, the time along AKC will exceed, by as much more, the time along ABC. Hence it appears that the time along ABC is the shortest possible; which was to be proven.

Reading 10
Newton: New Theory About Light and Colour

A Letter of Mr. Isaac Newton, *Professor of the Mathematicks in the University of Cambridge, containing his New Theory about Light and Colors: sent by the Author to the Publisher from Cambridge, Febr. 6.* 1671/72; *in order to be communicated to the* R. Society.

Sir,

To perform my late promise to you, I shall without further ceremony acquaint you that in the beginning of the year 1666 (at which time I applied myself to the grinding of optic glasses of other figures than spherical) I procured me a triangular glass prism to try therewith the celebrated phenomena of colours. And in order thereto having darkened my chamber and made a small hole in my window-shuts to let in a convenient quantity of the sun's light, I placed my prism at his entrance that it might be thereby refracted to the opposite wall. It was at first a very pleasing divertissement to view the vivid and intense colours produced thereby; but after a while, applying myself to consider them more circumspectly, I became surprised to see them in an oblong form, which according to the received laws of refraction I expected should have been circular.

They were terminated at the sides with straight lines, but at the ends the decay of light was so gradual that it was difficult to determine justly what was their figure; yet they seemed semicircular.

Comparing the length of this coloured spectrum with its breadth, I found it about five times greater, a disproportion so extravagant that it excited me to a more than ordinary curiosity of examining from whence it might proceed. I could scarce think that the various thickness of the glass or the termination with shadow or darkness could have any influence on light to produce such an effect; yet I thought it not amiss first to examine those circumstances, and so tried what would happen by transmitting light through parts of the glass of divers thicknesses, or through holes in the window of divers bignesses, or by setting the prism without, so that the light might pass through it and be refracted before it was terminated by the hole. But I found none of those circumstances material. The fashion of the colours was in all, these cases the same.

Then I suspected whether by any unevenness in the glass or other contingent irregularity these colours might be thus dilated. And to try this, I took another prism like the former and so placed it that the light, passing through them both, might be refracted contrary ways, and so by the latter returned into that course from which the former had diverted it. For by this means I thought the regular effects of the first prism would be destroyed by the second prism but the irregular ones more augmented by the multiplicity of refractions. The event was that the light which by the first prism was diffused into an oblong form was by the second reduced into an orbicular one with as much regularity as when it did not at all pass through them. So that, whatever was the cause of that length, 'twas not any contingent irregularity.

I then proceeded to examin[sic] more critically, what might be effected by the difference of the incidence of Rays coming from divsers parts of the Sun; and to that end, measured the several lines and angles, belonging to the Image. Its distance from the hole or Prisme was 22 foot; its utmost length 13¼ inches; its breadth 2⅝; the diameter of the hole ¼ of an inch; the angle, with the Rays, tending towards the middle of the image, made with those lines, in which they would have proceeded without refraction, was 44 deg.56'. And the vertical Angle of the Prisme, 63 deg. 12'. Also the Refractions on both sides the Prisme, that is, of the Incident, and Emergent Rays, were as near, as I could make them, equal, and consequently about 54 deg. 4'. And the Rays fell perpendicularly upon the wall. Now subducting the diameter of the hole from the lingth and breadth of the Image, there remains 13 Inches the length, and 2⅜ the breadth, comprehended by those Rays, which passed through the center of the said hole, and consequently the angel of the hole, which that breadth subtended, was about 31', answerable to the Suns Diameter; but the angle, which its length subtended, was more then five such diameters, namely 2 deg. 49'.

Having made these observations, I first computed from them the refractive power of that glass and found it measured by the ratio of the sines 20 to 31; and then by that ratio I computed the refractions of two rays flowing from opposite parts of the sun's *discus*, so as to differ 31 min. in their obliquity of incidence, and found that the emergent rays should have comprehended an angle of about 31 min. as they did before they were incident.

But because this computation was founded on the hypothesis of the proportionality of the sines of incidence and refraction, which though by my own experience I could not imagine to be so erroneous, as to make that angle but 31 min. which in reality was 2 deg. 49 min. yet my curiosity caused me again to take my prism: and having placed it at my window, as before, I observed, that by turning it a little about its axis to and fro, so as to vary its obliquity to the light, more than an angle of 4 or 5 degrees, the colours were not thereby sensibly translated from their place on the wall; and consequently by that variation of incidence, the quantity of refraction was not sensibly varied. By this experiment, therefore, as well as by the former computation, it was evident, that the difference of the incidence of rays, flowing from divers parts of the sun, could not make them after decussation diverge at a sensibly greater angle, than that which they before converged; which being at most but about 31 or 32 min. there still remained some other cause to be found out, from whence it could be 2 deg. 49 min.

Then I began to suspect, whether the rays, after their trajection through the prism, did not move in curved lines, and according to their more or less curvity, tend to divers parts of the wall. And it increased my suspicion, when I remember that I had often seen a tennis-ball, struck with and oblique racket, describe such a curved line. For, a circular as well as a progressive motion being communicated to it by that stroke, its parts, on that side where the motions conspire, must press and beat the contiguous air more violently that on the other , and there excite a reluctancy and reaction of the air proportionably greater. And for the same reason, if the rays of light should possibly be globular bodies, and by their oblique passage out of one medium into another acquire a circulating motion; they ought to feel the greater resistance from the ambient aether, on that side where the motions conspire, and thence be continually bowed to the other. But notwithstanding this plausible ground of suspicion, when I came to examine it, I could observe no such curvity in them. And besides (which was enough for my purpose) I observed, that the difference betwixt the length of the image and the diameter of the hole, through which the light was transmitted, was proportionable to their distance.

The gradual removal of these suspicions at length led me to the experimentum crucis, which was this; I took two boards, and placed one of them close behind the prism at the window, so that the light might pass through a small hole made in it for the purpose and fall on the other board, which I placed at about 12 feet distance, having first made a small hole in it also, for some of that incident light to pass through. Then I placed another prism behind this second board so that the light, targeted through both the boards, might pass through that also, and be again refracted before it arrived at the wall. This done, I took the first prism in my hand, and turned it to and fro slowly about its axis, so much as to make the several parts of the image cast on the second board successively pass through the hole in it, that I might observe to what places on the wall the second prism would refract them. And I saw by the variation of those places that the light tending to that end of the image towards which the refraction of the first prism was made did in the second prism suffer a refraction considerably greater than the light tending to the other end. And so the true cause of the length of that image was detected to be no other than that light consists of rays differently refrangible, which, without any respect to a difference in their incidence, were, according to their degrees of refrangibility, transmitted towards divers parts of the wall.

When I understood this, I left off my aforesaid glassworks; for I saw, that the perfection of the telescopes was hitherto limited, no so much for want of glasses truly figured according to the prescriptions of optick authors (which all men have hitherto imagined) as because that *light* itself is an heterogeneous mixture of *differently regrangible rays:* so that were a glass so exactly figured as to collect any on sort of rays into one point, it could not collect those also into the same point, which, having the same incidence upon the same medium, are apt to suffer a different refraction. Nay, I wondered, that seeing the difference of refrangivility was so great as I found, telescopes should arrive to that perfection they are now at: for, measuring the refractions in one of my prisms, I found, that supposing the common sine of incidence upon on of its planes was 44 parts, the sine of refraction of the umost rays on the red end of the colours, made out of the glass into the air, would be 68 parts; and the sine of refraction of the utmost rays on the other end, 69 patrs; so that the difference is about a 24th or 25th part of the whole refraction. And, consequently, the object-glass of any telescope cannot collect all the rays which come from one point of an object, so as to make them convene at its focus in less room than in a circular space, whose diameter is the 50th part of the diameter of its aperture; which is an irregularity, some hundreds of times greater than a circularly figured *lens*, of so small a section as the object-glasses of long telescopes are, would cause by the unfitness of its figure, were light uniform.

This made me take reflections into consideration; and finding them regular, so that the angle of reflection of all sorts of rays was equal to their angle of incidence, I understood, that by their mediation optick instruments might be brought to any degree of perfection imaginable provided a reflecting substance could be found which would polish as finely as glass, and reflect as much light as glass transmits, and the art of communicating to it a parabolick figure by also attained. But these seemed very great difficulties, and I have also thought them insuperable, when I farther considered, that every irregularity in a reflecting superficies makes the rays stray 5 or 6 times more out of their due course, than the like irregularities in a refracting one: so that a much greater curiosity would be here requisite, than in figuring glasses for refraction.

Amidst these thoughts, I was forced from *Cambridge, anno* 1666, by the intervening plague, and it was more than two years before I proceeded further. But then having thought on the tender way of polishing, proper for metal, whereby, as I imagined, the figure would be corrected to the last; I began to try what might be effected in this kind, and by degrees so far perfected an instrument in the essential parts of it like that I sent to *London*, by which I could discern *Jupiter's* four concomitants, and shewed them divers times to two other of my acquaintance. I could also discern the moon-like *phase* of *Venus*, but not very distinctly, nor without some niceness in disposing the instrument.

From that time I was interrupted till this last autumn, when I made another. And as that was sensibly better than the first, especially for day-objects, so I doubt not but they will be still brought to a much greater perfection by their endeavours, who, as you inform me, are taking care about it at *London*.

I have sometimes thought to make a microscope, which should have, instead of an object-glass, a reflecting piece of metal. For these instruments seem as capable of improvement as telescopes, and perhaps more; because but one reflective piece of metal is requisite in them, as you may perceive by the diagram; where AC representeth the object-metal; CD, the eye-glass; F, their common focus; and O, the other focus of the metal. In which the object is placed.

But to return from this digression, I told you, that light it not similar or homogeneral, but consists of difform rays, some of which are more refrangible than others: so that of those which are alike incident on the same medium, some shall be more refracted than other; and that not by virtue of the glass, or other external cause, but from a predisposition, which ever particular ray hath to suffer refraction.

I shall now proceed to acquaint you with another more notable difformity in its rays, wherein the origin of colours is unfolded: concerning which I shall lay down the doctrine first and then for its examination give you an instance or two of the experiments, as a specimen of the rest.

The doctrine you will find comprehended and illustrated in the following propositions.

1. As the rays of light differ in degrees of refrangibility, so they also differ in their disposition to exhibit this or that particular colour. Colours are not qualifications of light, derived from refractions or reflections of natural bodies (as 'tis generally believed), but original and connate properties which in divers rays are divers. Some rays are disposed to exhibit a red colour and no other, some a yellow and no other, some a green and no other, and so of the rest. Nor are there only rays proper and particular to the more eminent colours, but even to all their intermediate gradations.

2. To the same degree of refrangibility ever belongs the same colour, and to the same colour ever belongs the same degree of refrangibility. The least refrangible rays are all disposed to exhibit a red colour, and contrarily those rays which are disposed to exhibit a red colour are all the least refrangible. So the most refrangible rays are all disposed to exhibit a deep violet colour, and contrarily those which are apt to exhibit such a violet colour are all the most refrangible. And so to all the intermediate colours in a continued series belong intermediate degrees of refrangibility. And this analogy 'twixt colours and refrangibility is very precise and strict; the rays always either exactly agreeing in both or proportionally disagreeing in both.

3. The species of colour and degree of refrangibility proper to any particular sort of rays is not mutable by refraction nor by reflection from natural bodies nor by any other cause that I could yet observe. When any one sort of rays hath been well parted from those of other kinds, it hath afterwards obstinately retained its colour, notwithstanding my utmost endeavours to change it. I have refracted it with prisms and reflected it with bodies which in daylight were of other colours; I have intercepted it with the coloured film of air interceding two compressed plates of glass; transmitted it through coloured mediums and through mediums irradiated with other sorts of rays, and diversely terminated it; and, yet could never produce any new colour out of it. It would by contracting or dilating become more brisk or faint and by the loss of many rays in some cases very obscure and dark; but I could never see it changed in specie.

4. Yet seeming transmutations of colours may be made, where there is any mixture of divers sorts of rays. For in such mixtures, the component colours appear not, but by their mutual allaying each other constitute a middling colour. And therefore if by refraction or any other of the aforesaid causes the difform rays latent in such a mixture be separated, there shall emerge colours different from the colour of the composition. Which colours are not new generated, but only made apparent by being parted; for if they be again entirely mixed and blended together, they will again compose

that colour which they did before separation. And for the same reason, transmutations made by the convening of divers colours are not real; for when the difform rays are again severed, they will exhibit the very same colours which they did before they entered the composition—as you see blue and yellow powders when finely mixed appear to the naked eye green, and yet the colours of the component corpuscles are not thereby transmuted, but only blended. For, when viewed with a good microscope, they still appear blue and yellow interspersedly.

5. There are therefore two sorts of colours: the one original and simple, the other compounded of these. The original or primary colours are red, yellow, green, blue, and a violet-purple, together with orange, indigo, and an indefinite variety of intermediate graduations.

6. The same colours in specie with these primary ones may be also produced by composition. For a mixture of yellow and blue makes green; of red and yellow makes orange; of orange and yellowish green makes yellow. And in general if any two colours be mixed which, in the series of those generated by the prism, are not too far distant one from another, they by their mutual alloy compound that colour which in the said series appeareth in the mid-way between them. But those which are situated at too great a distance, do not so. Orange and indigo produce not the intermediate green, nor scarlet and green the intermediate yellow.

7. But the most surprising and wonderful composition was that of whiteness. There is no one sort of rays which alone can exhibit this. 'Tis ever compounded, and to its composition are requisite all the aforesaid primary colours, mixed in a due proportion. I have often with admiration beheld that, all the colours of the prism being made to converge and thereby to be again mixed as they were in the light before it was incident upon the, prism, reproduced light, entirely and perfectly white, and not at all sensibly differing from a direct light of the sun, unless when the glasses I used were not sufficiently clear; for then they would a little incline it to their colour.

8. Hence therefore it comes to pass that whiteness is the usual colour of light, for light is a confused aggregate of rays endued with all sorts of colours, as they are promiscuously darted from the various parts of luminous bodies. And of such a confused aggregate, as I said, is generated whiteness, if there be a due proportion of the ingredients; but if any one predominate, the light must incline to that colour, as it happens in the blue flame of brimstone, the yellow flame of a candle, and the various colours of the fixed stars.

9. These things considered, the manner how colours are produced by the prism is evident. For of the rays constituting the incident light, since those which differ in colour proportionally differ in infrangibility, they by their unequal refractions must be severed and dispersed into an oblong form in an orderly succession from the least refracted scarlet to the most refracted violet. And for the same reason it is that objects, when looked upon through a prism, appear coloured. For the difform rays, by their unequal refractions, are made to diverge towards several parts of the retina, and there express the images of things coloured, as in the former case they did the sun's image upon a wall.

And by this inequality of refractions they become not only coloured, but also very confused and indistinct.

10. Why the colours of the rainbow appear in falling drops of rain is also from hence evident. For those drops which refract the rays disposed to appear purple in greatest quantity to the spectator's eye, refract the rays of other sorts so much less as to make them pass beside it; and such are the drops on the inside of the primary bow and on the outside of the secondary or exterior one. So those drops which refract in greatest plenty the rays apt to appear red toward the spectator's eye, refract those of other sorts so much more as to make them pass beside it; and such are the drops on the exterior part of the primary and interior part of the secondary bow.

11. The odd phaenomena of an infusion of *lignum nephriticum*, leaf-gold, fragments of coloured glass, and some other transparently coloured bodies, appearing in one position of one colour, and of another in another, are on these grounds no longer riddles. For those are substances apt to reflect on sort of light, and transmit another; as may be seen in a dark room, by illuminating them with similar or uncompounded light. For then they appear of that colour only with which they are illuminated; but yet in one position more vivid and luminous than in another, accordingly as they are disposed more of less to reflect or transmit that incident colour.

12. From hence also is manifest the reason of an unexpected experiment, which Mr. Hook, somewhere in this *Micrography*, relates to have made with two wedge-like transparent vessels, filled the one with a Red, the other with a Blue liquor: namely, that though they were severally transparent enough, yet both together became opake; for if one transmitted only Red, and the other only blue, no rays could pass through both.

13. I might add more instances of this nature, but I shall conclude with this general one, that the colours of all natural bodies have no other origin than this, that they are variously qualified to reflect one sort of light in greater plenty than another. And this I have experimented in a dark room by illuminating those bodies with uncompounded light of divers colours. For by that means any body may be made to appear of any colour. They have there no appropriate colour, but ever appear of the colour of the light cast upon them, but yet with this difference, that they are most brisk and vivid in the light of their own daylight colour. Minium appeareth there of any colour indifferently with which 'tis illustrated, but yet most luminous in red, and so Bise appeareth indifferently of any colour with which 'tis illustrated, but yet most luminous in blue. And therefore minium reflecteth rays of any colour, but most copiously those endued with red; and consequently when illustrated with daylight, that is, with all sorts of rays promiscuously blended, those qualified with red shall abound most in the reflected light, and by their prevalence cause it to appear of that colour. And for the same reason bise, reflecting blue most copiously, shall appear blue by the excess of those rays in its reflected light; and the like of other bodies. And that this is the entire and adequate cause of their colours is manifest, because they have no power to change or alter the colours of any sort of rays incident apart, but put on all colours indifferently with which they are enlightened.

These things being so it can no longer be disputed whether there be colours in the dark, nor whether they be the qualities of the objects we see, no, nor perhaps whether light be a body. For since colours are the qualities of light, having its rays for their entire and immediate subject, how can we think those rays qualities also, unless one quality may be the subject of and sustain another—which in effect is to call it substance. We should not know bodies for substances were it not for their sensible qualities, and the principal of those being now found due to something else, we have as good reason to believe that to be a substance also.

Besides, who ever thought any quality to be a heterogeneous aggregate, such as light is discovered to be? But to determine more absolutely what light is, after what manner refracted, and by what modes or actions it produceth in our minds the phantasms of colours, is not so easy. And I shall not mingle conjectures with certainties.

Reviewing what I have written, I see the discourse itself will lead to divers experiments sufficient for its examination: and therefore I shall not trouble you farther than to describe one of those which I have already insinuated.

In a darkened room make a hole in the shut of a window, whose diameter may conveniently be about a third part of an inch, to admit a convenient quantity 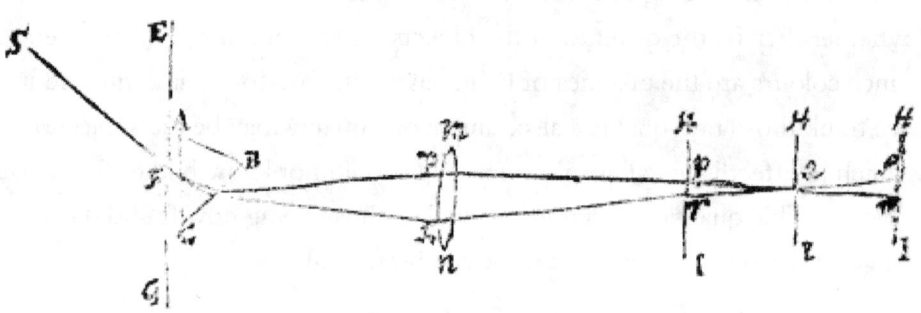 of the sun's light: and there place a clear and colourless prism, to refract the entering light towards the further part of the room; which as I said will thereby be diffused in to an oblong coloured image. Then place a lens of about 3 foot radius, suppose a broad object-glass of a three foot telescope at the distance of about 4 or 5 foot from thence, through which all those colours may at once be transmitted and made by its refraction to convene at a further distance of about 10 or 12 feet. If at that distance you intercept this light with a sheet of white paper, you will see the colours converted into whiteness again by being mingled. But it is requisite that the prism and lens be placed steady, and that the paper, on which the colours are cast, be moved to and fro; for by such motion you will not only find at what distance the Whiteness is most perfect, but also see how the colours gradually convene and vanish into whiteness; and afterwards, having crossed on another in that place where the compound whiteness, are again dissipated and sever, and, in an inverted order, retain the same colours which they had before they entered the composition. You may also see, that if any of the colours at the lens be intercepted, the whiteness will be changed into the other colours. And therefore, that the composition of whiteness be perfect, care must be taken that none of the colours fall besides the lens. In the annexed design of this experiment, ABC expresseth the prism set endwise to sight, close by the hole F, of the window EG. Its vertical angle, ACB, may conveniently be about 63 deg. MN designeth the lens: its breadth, 2½ or 3 inches. OF, one of the straight lines, in which difform rays may be conceived to flow successively from the sun. FP and FT, two of those rays unequally refracted, which the lens makes to converge towards X, and after decussation to diverge again. And PT, X, π, the paper at divers distances, on which h the colours are projected; which in X constitute whiteness, but are Red and Yellow in T and τ and Blue and Purple in P and π.

If you proceed further to try the impossibility of changing and unpomounded colour, which I have asserted in the 3d and 13th Propositions, it is requisite that the room be made very dark; lest any scattering light, mixing with the colour, disturb and allay it, and render it compound contrary to the design of the experiment: it is also requisite that there be a perfecter separation of the colours, than after the manner above described can be made by the refraction of one single prism; and how to make such further separations, will scarce be difficult to them that consider the discovered laws of refractions. But if trials shall be made with colours not throughly separated, there must be allowed changes proportionable to the mixture. Thus if compound Yellow light fall upon the blue bise, the bise will not appear perfectly tallow, but rather Green: because there are in the Yellow micture many rays endued with Green; and Green, being less remote from the usual Blue colour of the bise than Yellow, is the more copiously reflected by it.

In like manner, if any one of the prismatic colours, suppose Red, be intercepted, on design to try the asserted impossibility of reproducing the colour, either that the colours be very well parted before the Red be intercepted, or that together with the Red, the neighbouring colours, into which any Red is secretly dispersed, that is, the Yellow and perhaps Green too, be intercepted: or else, that allowance be made for the emerging of so much Red out of the Yellow-green, as my possible have been diffused, and scatteringly blended in those colours. And if these things be observed, the new production of Red, or of any other intercepted colour, will be found impossible. This I conceive is enough for introduction to experiments of this kind; which, if any of the Royal Society shall be so curious as to prosecute, I shall be very glad to be informed with what success: that if an thing seem to be defective, or to thwart this relation, I may have an opportunity of giving further direction about it; or of acknowledging my errors, if I have committed any.

Reading 11
Young: Lecture XXXIX. On the Nature of Light and Colors

3070 THE nature of light is a subject of no material importance to the concerns of life or to the practice of the arts, but it is in many other respects extremely interesting, especially as it tends to assist our views both of the nature of our sensations, and of the constitution of the universe at large. The examination of the production of colours, in a variety of circumstances, is intimately connected with the theory of their essential properties, and their causes; and we shall find that many of these
3075 phenomena will afford us considerable assistance in forming our opinion respecting the nature and origin of light in general.

It is allowed on all sides, that light either consists in the emission of very minute particles from luminous substances, which are actually projected, and continue to move, with the velocity commonly attributed to light, or in the excitation of an undulatory motion, analogous to that which
3080 constitutes sound, in a highly light and elastic medium pervading the universe; but the judgments of philosophers of all ages have been much divided with respect to the preference of one or the other of these opinions. There are also some circumstances which induce those, who entertain the first hypothesis, either to believe, with Newton, that the emanation of the particles of light is always attended by the undulations of an etherial medium, accompanying it in its passage, or to suppose,
3085 with Boscovich, that the minute particles of light themselves receive, at the time of their emission, certain rotatory and vibratory motions, which they retain as long as their projectile motion continues. These additional suppositions, however necessary they may have been thought for explaining some particular phenomena, have never been very generally understood or admitted, although no attempt has been made to accommodate the theory in any other manner to those
3090 phenomena.

We shall proceed to examine in detail the manner in which the two principal hypotheses respecting light may be applied to its various properties and affections; and in the first place to the simple propagation of light in right lines through a vacuum, or a very rare homogeneous medium. In this circumstance there is nothing inconsistent with either hypothesis; but it undergoes some
3095 modifications, which require to be noticed, when a portion of light is admitted through an aperture, and spreads itself in a slight degree in every direction. In this case it is maintained by Newton that the margin of the aperture possesses an attractive force, which is capable of inflecting the rays: but there is some improbability in supposing that bodies of different forms and of various refractive powers should possess an equal force of inflection, as they appear to do in the production of these
3100 effects; and there is reason to conclude from experiments, that such a force, if it existed, must extend to a very considerable distance from the surfaces concerned, at least a quarter of an inch, and perhaps much more,- which is a condition not easily reconciled with other phenomena. In the Huygenian system of undulation, this divergence or diffraction is illustrated by a comparison with the motions of waves of water and of sound, both of which diverge when they are ad mitted into a
3105 wide space through an aperture, so much indeed that it has usually been considered as an objection to this opinion, that the rays of light do not diverge in the degree that would be expected if they were analogous to the waves of water. But as it has been remarked by Newton, that the pulses of

sound diverge less than the waves of water, so it may fairly be inferred, that in a still more highly elastic medium, the undulations, constituting light, must diverge much less considerably than either.

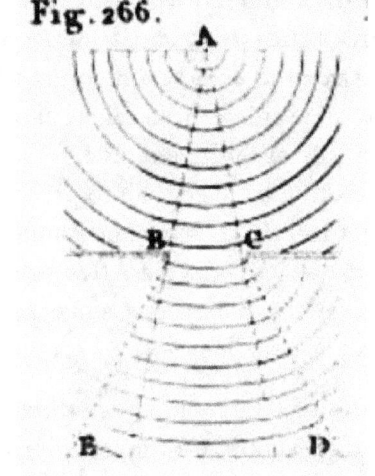

Fig. 266.

With respect, however, to the transmission of light through perfectly transparent mediums of considerable density, the system of emanation labours under some difficulties. It is not to be supposed that the particles of light can perforate with freedom the ultimate atoms of matter, which compose a 'substance of any kind; they must, therefore, be admitted in all directions through the pores or interstices of those atoms: for if we allow such suppositions as Boscovich's, that matter itself is penetrable, that is, immaterial, it is almost useless to argue the question further. It is certain that some substances retain all their properties when they are reduced to the thickness of the ten millionth of an inch at most, anti we cannot therefore suppose the distances of the atoms of matter in general to be so great as the hundred millionth of an inch. Now if ten feet of the most transparent water transmits, without interruption, one half of the light that enters it, each section or stratum of the thickness of one of these pores of matter must intercept only about one twenty thousand millionth, and so much must the space or area occupied by the particles be smaller than the interstices between them, and the diameter of each atom must be less than the hundred and forty thousandth part of its distance from the neighbouring particles: so that the whole space occupied by the substance must be as little filled, as the whole of England would be filled by a hundred men, placed at the distance of about thirty miles from each other. This astonishing degree of porosity is not indeed absolutely inadmissible, and there are many reasons for believing the statement to agree in some measure with the actual constitution of material sub stances; but the Huygenian hypothesis does not require the disproportion to be by any means so great, since the general direction and even the intensity of an undulation would be very little affected by the interposition of the atoms of matter, while these atoms may at the same time be supposed to assist in the transmission of the impulse, by propagating it through their own substance. Euler indeed imagined that the undulations of light might be transmitted through the gross substance of material bodies alone, precisely in the same manner as sound is propagated; but this supposition is for many reasons inadmissible.

A very striking circumstance, respecting the propagation of light, is the uniformity of its velocity in the same medium. According to the projectile hypothesis, the force employed in the free emission of light must be about a million million times as great as the force of gravity at the earth's surface; and it must either act with equal intensity on all the particles of light, or must impel some of them through a greater space than others, if its action be less powerful, since the velocity is the same in all cases; for example, if the projectile force is weaker with respect to red light than with respect to violet light, it must continue its action on the red rays to a greater distance than on the violet rays. There is no instance in nature besides of a simple projectile moving with a velocity uniform in all cases, whatever may be its cause, -and it is extremely difficult to imagine that so immense a force of repulsion can reside in all substances capable of becoming luminous, so that the light of decaying wood, or of two pebbles rubbed together, may be projected precisely with the same velocity, as the light emitted by iron burning in oxygen gas, or by the reservoir of liquid fire on the surface of the sun. Another cause would also naturally interfere with the uniformity of the velocity of light, if it consisted merely in the motion of projected corpuscles of matter; Mr. Laplace has calculated, that if any of the stars were 250 times as great in diameter as the sun, its attraction would be so strong as to destroy the whole momentum of the corpuscles of light proceeding from it, and to render the star invisible at a great distance; and although there is no reason to imagine that any of the stars are actually of this magnitude, yet some of them are probably many times greater than our sun, and therefore large enough to produce such a retardation in the motion of their light as would materially alter its effects. It is almost unnecessary to observe that the uniformity of the velocity of light, in those spaces which are free from all material substances, is a necessary consequence of the Huygenian hypothesis, since the undulations of every homogeneous elastic medium are always propagated, like those of sound, with the same velocity, as long as the medium remains unaltered.

On either supposition, there is no difficulty in explaining the equality of the angles of incidence and reflection; for these angles are equal as well in the collision of common elastic bodies with others incomparably larger, as in the reflections of the waves of water and of the undulations of sound. And it is equally easy to demonstrate, that the sines of the angles of incidence and refraction must be always in the same proportion at the same surface, whether it be supposed to possess an attractive force, capable of acting on the particles of light, or to be the limit of a medium through which the undulations are propagated with a diminished velocity. There are, however, some cases of the production of colours, which lead us to suppose that the velocity of light must be smaller in a denser than in a rarer medium; and supposing this fact to be fully established, the existence of such an attractive force could no longer be allowed, nor could the system of emanation be maintained by any one.

The partial reflection from all refracting surfaces is supposed by Newton to arise from certain periodical retardations of the particles of light, caused by undulations, propagated in all cases through an ethereal medium. The mechanism of these supposed undulations is so complicated, and attended by so many difficulties, that the few who have examined them have been in general entirely dissatisfied with them: and the internal vibrations of the particles of -light themselves, which Boscovich has imagined, appear scarcely to require a serious discussion. It may, therefore, safely be asserted, that in the projectile hypothesis this separation of the rays of light of the same kind by a partial reflection at every refracting surface, remains wholly unexplained. In the undulatory system, on the contrary, this separation follows as a necessary consequence. It is simplest to consider the ethereal medium which pervades any transparent substance, together with the material atoms of the substance, as constituting together a compound medium denser than the pure ether, but not more elastic; and by comparing the contiguous particles of the rarer and the denser medium with common elastic bodies of different dimensions, we may easily determine not only in what manner, but almost in what degree, this reflection must take place in different circumstances. Thus, if one of two equal bodies strikes the other, it communicates to it its whole motion without any reflection; but a smaller body striking a larger one is reflected, with the more force as the difference of their magnitude is greater; and a larger body, striking a smaller one, still proceeds with a diminished velocity; the remaining motion constituting, in the case of an undulation falling on a rarer medium, a part of a new series of motions which necessarily returns backwards with the appropriate velocity: and we may observe a circumstance nearly similar to this last in a portion of mercury spread out on a horizontal table; if a wave be excited at any part, it will be reflected from the termination of the mercury almost in the same manner as from a solid obstacle.

The total reflection of light, falling, with a certain obliquity, on the surface of a rarer medium, becomes, on both suppositions, a particular case of refraction. In the undulatory system, it is convenient to suppose the two mediums to be separated by a short space in which their densities approach by degrees to each other, in order that the undulation may be turned gradually round, so as to be reflected in an equal angle: but this supposition is not absolutely necessary, and the same effects may be expected at the surface of two mediums separated by an abrupt termination.

The chemical process of combustion may easily be imagined either to dis engage the particles of light from their various combinations, or to agitate the elastic medium by the intestine motions attending it: but the operation of Friction upon substances incapable of undergoing chemical changes, as well as the motions of the electric fluid through imperfect conductors, afford instances of the production of light in which there seems to be no easy way of supposing a decomposition of any kind. The phenomena of solar phosphori appear to resemble greatly the sympathetic sounds of musical instruments, which are agitated by other sounds conveyed to them through the air: it is difficult to understand in what state the corpuscles of light could he retained by these substances so as to be reemitted after a short space or time; and if it is true that diamonds are often found, which exhibit a red light after having received a violet light only, it-seems impossible to explain this property, on the supposition of the retention and subsequent emission of the same corpuscles.

The phenomena of the aberration of light agree perfectly well with the system of emanation; and if the ethereal medium, supposed to pervade the earth and its atmosphere, were carried along before it, and partook materially in its motions, these phenomena could not easily be reconciled with the theory of undulation. But there is no kind of necessity for such a supposition: it will not he denied by the advocates of the Newtonian opinion that all material bodies are sufficiently porous to leave a medium pervading them almost absolutely at rest ; and if this be granted, the effects of aberration will appear to be precisely the same in either hypothesis.

The unusual refraction of the Iceland spar has been most accurately and satisfactorily explained by Huygens, on the simple supposition that this crystal possesses the property of transmitting an impulse more rapidly in one direction than in another; whence he infers that the undulations constituting light must assume a spheroidical instead of a spherical form, and lays down such laws for the direction of its motion, as are incomparably more consistent with experiment than any attempts which have been made to accommodate the phenomena to other principles. It is true that nothing has yet been done to assist us in understanding the effects of a subsequent refraction by a second crystal, unless any person can be satisfied with the name of polarity assigned by Newton to a property which he attributes to the particles of light, and which he supposes to direct them in the species of refraction which they are to undergo: but on any hypothesis, until we discover the reason why a part of the light is at first refracted in the usual manner, and another part in the unusual manner, we have no right to expect that we should understand how these dispositions are continued or modified, when the process is repeated.

In order to explain, in the system of emanation, the dispersion of the rays of different colours by means of refraction, it is necessary to suppose that all refractive mediums have an elective attraction, acting more powerfully on the violet rays, in proportion to their mass, than on the red. But an elective attraction of this kind is a property foreign to mechanical philosophy, and when we use the term in chemistry, we only confess our incapacity to assign a mechanical cause for the effect, and refer to an analogy-with other facts, of which t-he intimate nature is perfectly unknown to us. Itis not indeed very easy to give a demonstrative theory of the dispersion of coloured light upon the supposition of undulatory motion; but we may derive a very satisfactory illustration from the well known effects of waves of different breadths. The simple calculation of the velocity of waves, propagated in a liquid perfectly elastic, or incompressible, and free from friction, assigns to them all precisely the same velocity, what ever their breadth may be: the compressibility of the fluids actually existing introduces, however, a necessity for a correction according to the breadth of the wave, and it is very easy to observe, in a river or a pond of consider» able depth, that the wider waves proceed much more rapidly than the narrower. We may, therefore, consider the pure ethereal medium as analogous to an infinitely elastic fluid, in which undulations of all kinds move with equal velocity, and material transparent substances, on the contrary, as resembling those fluids, in which we see the large waves advance beyond the smaller; and by supposing the red light to consist of larger or wider undulations and the violet of smaller, we may sufficiently elucidate the greater refrangibility of the red than of the violet light.

It is not, however, merely on the ground of this analogy that we may be induced to suppose the undulations constituting red light to be larger than those of violet light: a very extensive class of phenomena leads us still more directly to the same conclusion; they consist chiefly of the production of colours by means of transparent plates, and by diffraction or inflection, none of which have been explained, upon the supposition of emanation, in a manner sufficiently minute or comprehensive to satisfy the most candid even of the advocates for the projectile system; while on the other hand all of them may be at once understood, from the effect of the interference of double lights, in a manner nearly similar to that which constitutes in sound the sensation of a beat, when two strings, forming an imperfect unison, are heard to vibrate together.

Supposing the light of any given colour to consist of undulations, of a given breadth, or of a given frequency, it follows that these undulations must be liable to those effects which we have already examined in the case of the waves of water, and the pulses of sound. It has been shown that two equal series of waves, proceeding from centres near each other, may be seen to destroy each other's effects at certain points, and other points at to re double them; and the beating of two sounds has been explained from a similar interference. We are now to apply the same principles to the alternate union and extinction of colours.

In order that the effects of two portions of light may be thus combined, it is necessary that they be derived from the same origin, and that they arrive at the same point by different paths, in directions not much deviating from each other. This deviation may be produced in one or both of the portions by diffraction,

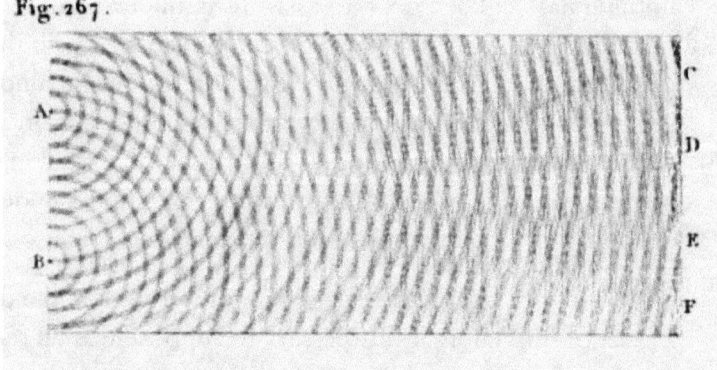

Fig. 267.

by reflection, by refraction, or by any of these effects combined; but the simplest case appears to be, when a beam of homogeneous light falls on a screen in which there are two very small holes or slits, which may be considered as centres of divergence, from whence the light is diffracted in every direction. In this case, when the two newly formed beams are received on a surface placed so as to intercept them, their light is divided by dark stripes into portions nearly equal, but becoming wider as the surface is more remote from the apertures, so as to subtend very nearly equal angles from the apertures at all distances, and wider also in the same proportion as the apertures are closer to each other. The middle of the two portions is always light, and the bright stripes on each side are at such distances, that the light, coming to them from one of the apertures, must have passed through a longer space than that which comes from the other, by an interval which is equal to the breadth of one, two, three, or more of the supposed undulations, while the intervening dark spaces correspond to a difference of half a supposed undulation, of one and a half, of two and a half, or more.

From a comparison of various experiments, it appears that the breadth of the undulations constituting the extreme red light must be supposed to be, in air, about one 86 thousandth of an inch, and those of the extreme violet about one 60 thousandth; the mean of the whole spectrum, with respect to the intensity of light, being about one 45 thousandth. From these dimensions it follows, calculating upon the known velocity of light, that almost 500 millions of millions of the slowest of such undulations must enter the eye in a single second. The combination of two portions of white or mixed light, when viewed at a great distance, exhibits a few white and black stripes, corresponding to this interval; although, upon closer inspection, the distinct effects of an infinite number of stripes of different breadths appear to be com pounded together, so as to produce a beautiful diversity of tints, passing by degrees into each other. The central whiteness is first changed to a yellowish, and then to a tawny colour, succeeded by crimson, and by violet and blue, which together appear, when seen at a distance, as a dark stripe; after this a green light appears, and the dark space beyond it has a crimson hue; the subsequent lights are all more or less green, the dark spaces purple and reddish; and the red light appears so far to predominate in all these effects, that the red or purple stripes occupy nearly the same place in the mixed fringes as if their light were received separately.

The comparison of the results of this theory with experiments fully establishes their general coincidence; it indicates, however, a slight correction in some of the measures, on account of some unknown cause, perhaps connected with the intimate nature of diffraction, which uniformly occasions the portions of light proceeding in a direction very nearly rectilinear, to be divided into stripes or fringes a little wider than the external stripes, formed by the light which is more bent.

Reading 12

Taylor: De motu Nervi tensi

Of the Motion of a Tense String, By Brook Taylor, Translated from the Latin.

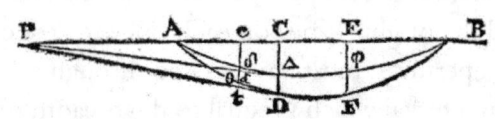

Lemma 1. — Let ADFB, and AΔφB, fig. 13, pl. 1, be two curves, so related that, drawing any ordinates CΔD, EφF, it is every where CΔ : CD :: Eφ : EF. Then the ordinates being diminished ad infinitum, so as the curves may coincide with the axis AB; I say that the ultimate ratio of the curvature in Δ to the curvature at D is as CΔ to CD.

Demonstration. — Draw the ordinate cδd very near to CD, and to D and Δ draw the tangents D*t* and Δθ, meeting the ordinate cd in *t* and θ. Then, because cδ: cd :: CΔ : CD, by hypothesis, the tangents produced will meet one another and the axis in the same point P. Hence by the similar triangles CDP, *ct*P, and CΔP, *c*θP, it will be *c*θ : *ct* :: CΔ : CD (:: *c*δ : *cd* by hypothesis) :: δθ (= *c*θ — *c*δ) : *dt* {= *ct* — *cd*). But the curvatures in Δ and D, are as the angles of contact θΔδ and *t*D*d*; and because δΔ and *d*D coincide with *c*C, those angles are as their subtenses δθ and *dt*, that is, by the analogy above, as CΔ and CD. Therefore, &c. Q. E. D.

Lemma 2. — At any instant of its vibration, let a tense cord, stretched between the points A and B, take any form of curve A*p*πB, fig. 14. Then will the increment of the velocity of any point P, or the acceleration arising from the force of tension in the string, be as the curvature of the string in the same point.

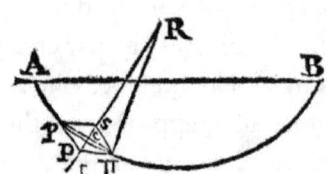

Demonstration. — Conceive the string to consist of equal rigid particles, infinitely small, as *p*P and Pπ; and erect the perpendicular PR = the radius of curvature at P, in which let the tangents *pt* and π*t* meet at *t* and their parallels π*s* and *ps* at *s*, also the chord *p*π in *c*. Then, by the principles of mechanics, the absolute force, by which the two particles *p*P and Pπ are urged towards R, will be to the force of tension in the string, as *st* to *pt*; and half this force, by which one particle *p*P is urged, will be to the tension of the string, as *ct* to *tp*, that is, because of the similar triangles *ctp*, *tp*R, as *tp* or P*p* to R*t* or PR. Therefore, because of the force of tension being given, the absolute accelerating force will be as P*p*/PR. But the acceleration generated is in a ratio composed of the ratios of the absolute force directly and of the matter moved inversely; and the matter moved being as the particle P*p*; therefore the acceleration is as 1/PR, that is, as the curvature at P. For the curvature is reciprocally as the radius of the osculatory circle, Q. E. D.

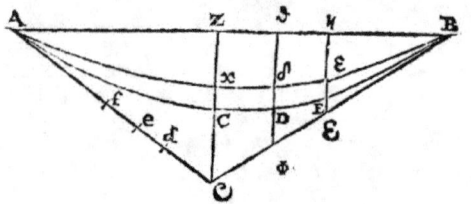

Prob. 1. *To determine the Motion of a Tense String.* — In this and the following problems, I suppose the string to move through a very small space from the axis of motion; and that the increment of tension

from the increase of the length, as also the obliquity of the radii of curvature, may be safely neglected.

Therefore let the string be stretched between the points A and B, fig. 15; and by a bow let the point z be drawn to the distance Cz from the axis AB. Then, taking away the bow, because of the flexure in the point c alone, that will first begin to move, by lemma 2. But as soon as the string is bent in the nearest points φ and d, these points will also begin to move; and then E and e; and so on. Also because of the great flexure in C, that point will at first move very swiftly; and thence the curvature being increased in the next points D, E, &c. these will be accelerated very swiftly, and at the same time the curvature in C being diminished, that point in its turn will be accelerated more slowly. And in general, those points which are slower being accelerated the more, and those that are quicker, less accelerated, it will be brought about at length, that the forces being duly tempered to each other, all the motions will conspire together, and all the points will at the same time approach the axis, going and returning alternately ad infinitum.

Now for this purpose, the string must assume the form of a curve ACDEB, the curvature of which, in any point E, is as its distance Eη from the axis; the velocities of the points C, D, E, &c. being also in the ratio of the distances from the axis, Cz, Dθ, Eη, &c. For in this case the spaces Cx, Dδ, Eε, &c. described in the same infinitely small time, will be to each other as the velocities, that is, as the spaces to be run through Cz, Dθ, &c. Therefore the remaining spaces, xz, $\delta\theta$, $\varepsilon\eta$, &c. will be to each other in the same ratio. Also, by lemma 2, the accelerations will be to each other in the same ratio. So that, the ratio of the velocities always continuing the same as the ratio of the spaces to be described, all the points will arrive at the axis together, and all at once depart from it; therefore the curve ACDEB is rightly determined, Q. E. D.

Further, the two curves ACDEB, A$x\delta\varepsilon$B being compared together, by lemma 1, the curvatures in D and d will be as the distances from the axis Dθ and $\delta\theta$; therefore, by lemma 2, the acceleration of any given point in the string, will be as its distance from the axis. Hence, by sect. 10, prop. 51, of Newton's Principia, all the vibrations, both great and small, will be performed in the same periodical time, and the motion of any point be similar to the oscillation of a body vibrating in a cycloid, Q. E. I.

Corol. — Curvatures being reciprocally as the radii of the osculating circles; therefore, let a denote a given line, then will its radius of curvature at E be $= aa/E\eta$.

Reading 13
Gilbert: De Magnete

WILLIAM GILBERT. BOOK FIRST. CHAPTER I.

WRITINGS OF ANCIENT AND MODERN AUTHORS CONCERNING THE LOADSTONE: VARIOUS OPINIONS AND DELUSIONS.

In former times when philosophy, still rude and uncultured, was involved in the murkiness of errors and ignorances, a few of the virtues and properties of things were, it is true, known and understood : in the world of plants and herbs all was confusion, mining was undeveloped, and mineralogy neglected. But when, by the genius and labors of many workers, certain things needful for man's use and welfare were brought to light and made known to others (reason and experience meanwhile adding a larger hope), then did mankind begin to search the forests, the plains, the mountains and precipices, the seas and the depths of the waters, and the inmost bowels of earth, and to investigate all things. And by good luck at last the loadstone was found, as seems probable, by iron-smelters or by miners in veins of iron ore. On being treated by the metallurgists, it quickly exhibited that strong powerful attraction of iron — no latent nor obscure property, but one easily seen of all; one observed and commended with many praises. And after it had come forth as it were out of darkness and out of deep dungeons and been honored of men on account of its strong and marvellous attraction of iron, then many ancient philosophers and physicians discoursed of it, and briefly (but briefly only) made it matter of record: as, for instance, Plato in the *Io*, Aristotle only in his first book *De Anima*; likewise Theophrastus the Lesbian, Dioscorides, Caius Plinius secundus, Julius Solinus. These record only that the loadstone attracts iron: its other properties were all hid. But lest the story of the loadstone should be jejune and too brief, to this one sole property then known were appended certain figments and falsehoods which in the early time no less than nowadays were by precocious sciolists and copyists dealt out to mankind to be swallowed. For example, they asserted that a loadstone rubbed with garlic does not attract iron; nor when it is in presence of a diamond. Lucretius Caruss, the Epicurean poet, deems the attraction to be due to this, that as there is from all things an efflux of minutest bodies, so there is from iron efflux of atoms into the space betwixt the iron and the loadstone — a space emptied of air by the loadstone's atoms (seeds); and when these begin to return to the loadstone, the iron follows, the corpuscles being entangled with each other. Something similar is said by Joannes Costaeus, following Plutarch. Thomas Aquinas, in his *Physica*, Bk. 7, treating briefly of the loadstone, gets at the nature of it fairly well: with his godlike and perspicacious mind he would have developed many a point had he been acquainted with magnetic experiments. Plato holds the magnetic virtue to be divine. But when, some three or four hundred years ago, the magnetic movement to the north and the south was discovered or recognized anew, many learned men, each according to his own gifts, strove to honor with admiration and praise or to explain with feeble reasonings a property so curious and so necessary for the use of mankind. Of more recent authors, very many have striven to discover the cause of this direction and movement to north and south, and to understand this so great miracle of nature and lay it open to others: but they wasted oil and labor, because, not being practical in the research of objects in nature, being acquaint only with books, being led astray by certain erroneous physical systems, and having made no magnetical experiments, they constructed certain raciocinations on a basis of mere opinions, and old-womanishly dreamt the things that were not.

CHAPTER III.

THE LOADSTONE POSSESSES PARTS DIFFERING IN THEIR NATURAL POWERS, AND HAS POLES CONSPICUOUS FOR THEIR PROPERTIES.

The many qualities exhibited by the loadstone itself qualities hitherto recognized yet not well investigated, are to be pointed out in the first place, to the end the student may understand the powers of the loadstone and of iron, and not be confused through want of knowledge at the threshold of the arguments and demonstrations. In the heavens, astronomers give to each moving sphere two poles; thus do we find two natural poles of excelling importance even in our terrestrial globe, constant points related to the movement of its daily revolution, to wit, one pole pointing to Arctos (Ursa) and the north ; the other looking toward the opposite part of the heavens. In like manner the loadstone has from nature its two poles, a northern and a southern; fixed, definite points in the stone, which are the primary termini of the movements and effects, and the limits and regulators of several actions and properties. It is to be understood, however that not from a mathematical point does the force of the stone emanate, but from the parts themselves ; and all these parts in the whole — while they belong to the whole — the nearer they are to the poles of the stone the stronger virtues do they acquire and pour out on other bodies. These poles look toward the poles of the earth, and move toward them, and are subject to them. The magnetic poles may be found in every loadstone, whether strong and powerful (male, as the term was in antiquity) or faint, weak, and female; whether its shape is due to design or to chance, and whether it be long, or flat, or four square, or three-cornered, or polished; whether it be rough, broken-off, or unpolished: the loadstone ever has and ever shows its poles. But inasmuch as the spherical form, which, too, is the most perfect, agrees best with the earth, which is a globe, and also is the form best suited for experimental uses, therefore we purpose to give our principal demonstrations with the aid of a globe-shaped loadstone, as being the best and most fitting. Take then a strong loadstone, solid, of convenient size, uniform, hard, without flaw; on a lathe, such as is used in turning crystals and some precious stones, or on any like instrument (as the nature and toughness of the stone may require, for often it is worked only with difficulty), give the loadstone the form of a ball. The stone thus prepared is a true homogeneous offspring of the earth and is of the same shape, having got from art the orbicular form that nature in the beginning gave to the earth, the common mother ; and it is a natural little body endowed with a multitude of properties whereby many abstruse and unheeded truths of philosophy, hid in deplorable darkness, may be more readily brought to the knowledge of mankind. To this round stone we give the name Μικρόγη (microge) or Terrella (earthkin, little earth).

To find, then, poles answering to the earth's poles, take in your hand the round stone, and lay on it a needle or a piece of iron wire : the ends of the wire move round their middle point, and suddenly come to a standstill Now, with ochre or with chalk, mark where the wire lies still and sticks. Then move the middle or centre of the wire to another spot, and so to a third and a fourth, always marking the stone along the length of the wire where it stands still: the lines so marked will exhibit meridian circles, or circles like meridians on the stone or terrella; and manifestly they will all come together at the poles of the stone. The circles being continued in this way, the poles appear, both the north and the south, and betwixt these, midway, we may draw a large circle for an equator, as is done by the astronomer in the heavens and on his spheres and by the geographer on the terrestrial globe; for the line so drawn on this our terrella is also of much utility in our demonstrations and our magnetic experiments. Poles are also found in the round stone, in a versorium, in a piece of iron touched with a loadstone and resting on a needle or point (attached at its base to the terrella), so that it can freely revolve, as in the figure.

On top of the stone AB is set the versorium in such a way that its pointer may remain in equilibrium: mark with chalk the direction of the pointer when at rest. Then move the instrument to another spot and again mark the direction in which the pointer looks; repeat this many times at many different points and you will, from the convergence of the lines of direction, find one pole at the point A, the other at B. A pointer also indicates the true pole if brought near to the stone, for it eagerly faces the stone at right angles, and seeks the pole itself direct and turns on its axis in a right line toward the centre of the stone. Thus the pointer D regards A and F, the pole and the centre, but the pointer E looks not straight either toward the pole A or the centre F. A bit of fine iron wire as long as a barley-corn is laid on the stone and is moved over the zones and the surface of the stone till it stands perpendicularly erect; for at the poles, whether N. or S., it stands erect ; but the farther it is from the poles (towards the equator) the more it inclines. The poles thus found, you are to mark with a sharp file or a gimlet.

CHAPTER IV.

WHICH POLE IS THE NORTH: HOW THE NORTH POLE IS DISTINGUISHED FROM THE SOUTH POLE.

One of the earth's poles is turned toward Cynosura and steadily regards a fixed point in the heavens (save that it is unmoved by the precession of the fixed stars in longitude, which movement we recognize in the earth, as we shall later show); the other pole is turned toward the opposite aspect of the heavens, an aspect unknown to the ancients, but which is adorned with a multitude of stars, and is itself a striking spectacle for those who make long voyages. So, too, the loadstone possesses the virtue and power of directing itself toward the north and the south (the earth itself co-operating and giving to it that power) according to the conformation of nature, which adjusts the movements of the stone to its true locations. In this manner it is demonstrated; Put the magnetic stone (after you have found the poles) in a round wooden vessel — a bowl or a dish; then put the vessel holding the magnet (like a boat with a sailor in it) in a tub of water or a cistern where it may float freely in the middle without touching the rim, and where the air is not stirred by winds (currents) which might interfere with the natural movement of the stone: there the stone, as if in a boat floating in the middle of an unruffled surface of still water, will straightway set itself, and the vessel containing it in motion, and will turn in a circle till its south pole shall face north and its north pole, south. For, from a contrary position, it returns to the poles; and though with its first too strong impetus it passes beyond, still, as it comes back again and again, at last it rests at the poles or in the meridian (save that, according to the place, it diverges a very little from those points, or from the meridional line, the cause of which we will define later). As often as you move it out of its place, so often, by reason of the extraordinary power with which nature has endowed it, does it seek again its fixed and determinate points. Nor does this occur only when the poles of the loadstone in the float are made to lie evenly in the plane of the horizon; it takes place also even though one pole, whether north or south, be raised or depressed lo, 20, 30, 40, or 80 degrees from the plane of the horizon ; you shall see the north part of the stone seek the south, and the south part the north ; so that if the pole of the stone be but one degree from the zenith and the centre of the heavens, the whole stone revolves until the pole finds its own place ; and though the pole does not point exactly to its seat, yet it will incline toward it, and will come to rest in the meridian of its true direction. And it moves with the same impetus whether the north pole be directed toward the upper heavens, or whether the south pole be raised above the horizon. Yet it must always be borne in mind that though there are manifold differences between stones, and one far surpasses another in virtue and efficiency, still all loadstones have the same limits and turn to the same points. Further, it is to be remembered that all who hitherto have written about the poles of the loadstone, all instrument-makers, and navigators, are egregiously mistaken in taking for the north pole of the loadstone 'the part of the stone that inclines to the north, and for the south pole the part that looks to the south : this we will hereafter prove to be an error. So ill-cultivated is the whole philosophy of the magnet still, even as regards its elementary principles.

CHAPTER V.

ONE LOADSTONE APPEARS TO ATTRACT ANOTHER IN THE NATURAL POSITION; BUT IN THE OPPOSITE POSITION REPELS IT AND BRINGS IT TO RIGHTS.

First we have to describe in popular language the potent and familiar properties of the stone; afterward, very many subtle properties, as yet recondite and unknown, being involved in obscurities, are to be unfolded ; and the causes of all these (nature's secrets being unlocked) are in their place to be demonstrated in fitting words and with the aid of apparatus. The fact is trite and familiar, that the loadstone attracts iron; in the same way, too, one loadstone attracts another. Take the stone on which you have designated the poles, N. and S., and put it in its vessel so that it may float; let the poles lie just in the plane of the horizon, or at least in a plane not very oblique to it; take in your hand another stone the poles of which are also known, and hold it so that its south pole shall lie toward the north pole of the floating stone, and near it alongside ; the floating loadstone will straightway follow the other (provided it be within the range and dominion of its powers), nor does it cease to move nor does it quit the other till it clings to it, unless, by moving your hand away, you manage skilfully to prevent the conjunction. In like manner, if you oppose the north pole of the stone in your hand to the south pole of the floating one, they come together and follow each other. For opposite poles attract opposite poles. But, now, if in the same way you present N. to N. or S. to S., one stone repels the other; and as though a helmsman were bearing on the rudder it is off like a vessel making all sail, nor stands nor stays as long as the other stone pursues. One stone also will range the other, turn the other around, bring it to right about and make it come to agreement with itself. But when the two come together and are conjoined in nature's order, they cohere firmly. For example, if you present the north pole of the stone in your hand to the Tropic of Capricorn (for so we may distinguish with mathematical circles the round stone or terrella, just as we do the globe itself) or to any point between the equator and the south pole: immediately the floating stone turns round and so places itself that its south pole touches the north pole of the other and is most closely joined to it. In the same way you will get like effect at the other side of the equator by presenting pole to pole ; and thus by art and contrivance we exhibit attraction and repulsion, and motion in a circle toward the concordant position, and the same movements to avoid hostile meetings. Furthermore, in one same stone we are thus able to demonstrate all this: but also we are able to show how the self-same part of one stone may by division become either north or south. Take the oblong stone *ad* in which *a* is the north pole and *d* the south. Cut the stone in two equal parts and put part *a* in a vessel and let it float in water.

You will find that *a*, the north point, will turn to the south as before; and in like manner the point *d* will move to the north, in the divided stone, as before division. But *b* and *c*, before connected, now separated from each other, are not what they were before, *b* is now south while c is north. *b* attracts *c*, longing for union and for restoration of the original continuity. They are two stones made out of one, and on that account the *c* of one turning toward the *b* of the other, they are mutually attracted, and, being freed from all impediments and from their own weight, borne as they are on the surface of the water, they come together and into conjunction. But if you bring the part or point *a* up to *c* of the other, they repel one another and turn away; for by such a position of the parts nature is crossed and the form of the stone is perverted: but nature observes strictly the laws it has imposed upon bodies: hence the flight of one part from the undue position of the other, and hence the discord unless everything is arranged exactly according to nature. And nature will not suffer an unjust and inequitable peace, or an unjust and inequitable peace and agreement, but makes war and employs force to make bodies acquiesce fairly and justly. Hence, when rightly arranged, the parts attract each other, i.e., both stones, the weaker and the stronger, come together and with all their might tend to union: a fact manifest in all loadstones, and not, as Pliny supposed, only in those from Ethiopia. The Ethiopic stones if strong, and those brought from China, which are all powerful stones, show the effect most quickly and most plainly, attract with most force in the parts nighest the pole, and keep turning till pole looks straight on pole. The pole of a stone has strongest attraction for that part of another stone which answers to it (the adverse as it is called); e.g., the north pole of one has strongest attraction for, has the most vigorous pull on, the south part of another: so too it attracts iron more powerfully, and iron clings to it more firmly, whether previously magnetized or not. Thus it has been settled by nature, not without reason, that the parts nigher the pole shall have the greatest attractive force; and that in the pole itself shall be the seat, the throne as it were, of a high and splendid power; and that magnetic bodies brought near thereto shall be attracted most powerfully and relinquished with most reluctance. So, too, the poles are readiest to spurn and drive away what is presented to them amiss, and what is inconformable and foreign.

Book V CHAPTER XII

THE MAGNETIC FORCE IS ANIMATE, OR IMITATES A SOUL; IN MANY RESPECTS IT SURPASSES THE HUMAN SOUL WHILE THAT IS UNITED TO AN ORGANIC BODY.

Wonderful is the loadstone shown in many experiments to be, and, as it were, animate. And this one eminent property is the same which the ancients held to be a soul in the heavens, in the globes, and in the stars, in sun and moon. For they deemed that not without a divine and animate nature could movements so diverse be produced, such vast bodies revolve in fixed times, or potencies so wonderful be infused into other bodies; whereby the whole world blooms with most beautiful diversity through this primary form of the globes themselves. The ancient philosophers, as Thales, Heraclitus, Anaxagoras, Archelaus, Pythagoras, Empedocles, Parmenides, Plato and all the Platonists — nor Greek philosophers alone, but also the Egyptian and Chaldean, — all seek in the world a certain universal soul, and declare the whole world to be endowed with a soul. Aristotle held that not the universe is animate, but the heavens only; his elements he made out to be inanimate; but the stars were for him animate. As for us, we find this soul only in the globes and in their homogenic parts, and albeit this soul is not in all globes the same (for that in the sun or in certain stars is much superior to that in other less noble globes). Still in very many globes the souls agree in their powers. Thus, each homogenic part tends to its own globe and inclines in the direction common to the whole world, and in all globes the effused forms reach out and are projected in a sphere all round, and have their own bounds — hence the order and regularity of all the motions and revolutions of the planets, and their circuits, not pathless, but fixed and determinate, wherefore Aristotle concedes to the spheres and heavenly orbs (which he imagines) a soul, for the reason that they are capable of circular motion and action and that they move in fixed, definite, tracks. And I wonder much why the globe of earth with its effluences should have been by him and his followers condemned and driven into exile and cast out of all the fair order of the glorious universe, as being brute and soulless. In comparison with the whole creation 'tis a mere mite, and amid the mighty host of many thousands is lowly, of small account, and deformate. And to it the Aristotelians add allied elements that by like ill-fortune are also beggarly and despicable. Thus Aristotle's world would seem to be a monstrous creation, in which all things are perfect, vigorous, animate, while the earth alone, luckless small fraction, is imperfect, dead, inanimate, and subject to decay. On the other hand, Hermes, Zoroaster, Orpheus, recognize a universal soul. As for us, we deem the whole world animate, and all globes, all stars, and this glorious earth, too, we hold to be from the beginning by their own destinate souls governed and from them also to have the impulse of self-preservation. Nor are the organs required for organic action lacking, whether implanted in the homogenic nature or scattered through the homogenic body, albeit these organs are not made up of viscera as animal organs are, nor consist of definite members; indeed in some plants and shrubs the organs are hardly recognizable, nor are visible organs essential for life in all cases. Neither in any of the stars, nor in the sun, nor in the planets that are most operant in the world, can organs be distinguished or imagined by us; nevertheless, they live and endow with life small bodies at the earth's elevated points. If there is aught of which man may boast, that of a surety is soul, is mind ; and the other animals, too, are ennobled by soul ; even God, by whose rod all things are governed, is soul. But who shall assign organs to the divine intellects, seeing that they are superior to all organ-structure, nor are comprised in material organs? But in the bodies of the several stars the inborn energy works in ways other than in that divine essence which presides over nature; and in the stars, the sources of all things, in other ways than in animals; finally, in animals in other ways than in plants. Pitiable is the state of the stars, abject the lot of earth, if this high dignity of soul is denied them, while it is granted to the worm, the ant, the roach, to plants and morels; for in that case worms, roaches, moths, were more beauteous objects in nature and more perfect, inasmuch as nothing is excellent, nor precious, nor eminent, that

3625 hath not soul. But since living bodies spring from earth and sun and by them are animate, and since in the earth herbage springs up without sowing of seeds (e.g., when soil is taken out of the bowels of the earth and carried to some great elevation or to the top of a lofty tower and there exposed to the sunshine, after a little while a miscellaneous herbage springs up in it unbidden), it is not likely that they (sun and earth) can do that which is not in themselves ; but they awaken souls, and
3630 consequently are themselves possessed of souls. Therefore the bodies of the globes, as being the foremost parts of the universe, to the end they might be in themselves and in their state endure, had need of souls to be conjoined to them, for else there were neither life, nor prime act, nor movement, nor unition, nor order, nor coherence, nor *conactus* nor *sympathia* nor any generation, nor alternation of seasons, and no propagation; but all were in confusion and the entire world lapse into chaos, and,
3635 in fine, the earth were void and dead and without any use. But only on the superficies of the globes is plainly seen the host of souls and of animate existences, and in their great and delightful diversity the Creator taketh pleasure. But the souls (in the interior of the globes) confined, as it were, by prison bars send not forth their effused immaterial forms beyond the limits of the body, nor are bodies put in motion by them without labor and exertion; a breath carries and bears them forth; but
3640 if that breath be fouled or stilled by mischance, the bodies lie like the world's recrement or as the waste matter of the globes. But the globes themselves remain and endure, rotate and move in orbits, and without wasting or weariness run their courses. The human soul uses reason, sees many things, investigates many more; but, however well equipped, it gets light and the beginnings of knowledge from the outer senses, as from beyond a barrier — hence the very many ignorances and
3645 foolishnesses whereby our judgments and our life-actions are confused, so that few or none do rightly and duly order their acts. But the earth's magnetic force and the formate soul or animate form of the globes, that are without senses, but without error and without the injuries of ills and diseases, exert an unending action, quick, definite, constant, directive, motive, imperant, harmonious, through the whole mass of matter; thereby are the generation and the ultimate decay of all things on
3650 the superficies propagated. For if it were not for the movement whereby the daily revolution is accomplished, all things here on earth were wild and disordered, and worse than desert and unused would they ever remain. Yet these movements in nature's founts are not produced by thoughts or reasonings or conjectures, like human acts, which are contingent, imperfect, and indeterminate, but connate in them are reason, knowledge, science, judgment, whence proceed acts positive and
3655 definite from the very foundations and beginnings of the world: these, because of the weakness (*imbecillitatem*) of our soul, we cannot comprehend. Wherefore, not without reason, Thales, as Aristotle reports in his book De Anima, declares the loadstone to be animate, a part of the animate mother earth and her beloved offspring.

Reading 14

Du Fay: A Letter from Mons. Du Fay, F. R. S. and of the Royal Academy of Sciences at Paris, to his Grace CHARLES Duke of Richmond and Lenox, concerning Electricity.

Translated from the French by T. S. *M D*.

Paris, December 17, 1733.

My LORD,

I flatter myself your Grace will not be displeased with an Account of some extraordinary Discoveries I have made in the Electricity of Bodies, nor refuse the Favour I have to ask, that it may be communicated to the Royal Society. I owe this Homage to that Illustrious Body, not only as a Member thereof, but in this respect as a Debtor to their Works; for the Writings of Mr. Gray, and the late Mr. Hauksheey both of that Society, first put me upon the Subject, and furnished me with the Hints that led me to the following Discoveries. First, I have found that all Bodies (metallick, soft or fluid ones excepted) may be made Electrick, by first heating them more or left, and then rubbing them on any fort of Cloth. So that all kinds of Stones, as well precious as common, all sorts of Wood, and in general every thing that I have made Trial of, became Electrick, by heating and rubbing; except such Bodies as grow soft by Heat, as the Gums, which dislolve in Water, Glue, and such other Substances. 'Tis also to be remarked, that the hardest Stones and Marbles require more chafing or heating than others, and that the fame Rule obtains with regard to the Woods; for that Box, *Lignum Vitae* and such others must be chafed almost to the Degree of burning, whereas Fir, Lime-Tree and Cork require but a moderate Heat.

Sixthly, On making the Experiment related by Otho de Guerik, in his Collection of Experiments *de Spatio Vacuo* which consists in making a Ball of Sulphur render'd Electrical, to repel a Down-Feather, I perceived that the fame Effects were produced not only by the Tube, but by all electrick Bodies whatsoever; and I discovered a very simple Principle, which accounts for a great Part of the Irregularities, and if I may use the Term, of the Caprices that seem to accompany moil of the Experiments on Electricity. This Principle is, that Electrick Bodies attract all those that are not so, and repel them as soon as they are become electrick, by the Vicinity or Contact of the electrick Body. Thus Leaf-Gold is first attracted by the Tube; and acquires an Electricity by approaching it; and of confequence is immediately repell'd by it. Nor is it re-attracted, while it retains its electrick Quality. But if, while it is thus sustain'd in the Air, it chance to light on some other Body, it straightways loses its Electricity; and consequently is re-attracted by the Tube, which, after having given it a new Electricity, repels it a second time; which continues as long as the Tube keeps its Electricity. Upon applying this Principle to the various Experiments of Electricity, one will be surprised at the Number of obscure and puzzling Facts it clears up. For Mr. Hanksbee's famous Experiment of the Glass Globe, in which Silk Threads are put, is a necessary Consequence of it. When these Threads are ranged in Form of Rays by the Electricity of the Sides of the Globe, if the Finger be put near the Outside of the Globe, the Silk Threads within fly from it, as is well known; which happens only because the Finger, or any other Body applied near the

Glass Globe, is thereby render'd electrical, and consequently repels the Silk Threads, which are endow'd with the like Quality. With a little Reflection one may in the fame manner account for most of the other Phanomena, and which seem inexplicable, without attending to this Principle.

Seventhly, Chance has thrown in my way another Principle, more universal and remarkable than the preceding one, and which casts a new Light on the Subject of Electricity. This Principle is, that there are two distinct Electricities, very different from one another; one of which I call vitreous electricity and the other resinous Electricity. The first is that of Glass, Rock-Crystal, Precious Stones, Hair of Animals, Wool, and many other Bodies: The second is that of Amber, Copal, Gum-Lack, Silk, Thread, Paper, and a vast Number of other Substances. The Characteristick of these two Electricities is, that a Body of the vitreous Electricity, for Example, repels all such as are of the fame Electricity; and on the contrary, attracts all those of the resinous Electricity so that the Tube, made electrical, will repel Glass, Crystal, Hair of Animals, &c. when rendered electrick and will attract Silk, Thread, Paper, &c. though render'd electrical likewise. Amber on the contrary will attract electrick Glass, .and other Substances of the same Class, and will repel Gum-Lac, Copal, Silk, Thread, &c. Two Silk Ribbons rendered electrical, will repel each other two Woollen Threads will do the like; but a Woollen Thread and a Silk Thread will mutually attract one another. This Principle very naturally explains, why the Ends of Threads, of Silk, or Wool, recede from one another in Form of a Pencil or Broom, when they have acquired an electrick Quality. From this Principle one may with the same Ease deduce the Explanation of a great Number of other Phenomena. And 'tis probable, that this Truth will lead us to the further Discovery of many other things. In order to know immediately, to which of the two Classes of Electricity belongs any Body whatsoever, one need only render Electrical a Silk Thread, which is known to be of the resinous Electricity and see whether that Body, render'd electrical, attracts or repels it. if it attracts, 'tis certainly of that kind of Electricity which I call vitreous; if on the contrary it repels, 'tis-of the same kind of Electricity with the Silk, that is, of the resinous. I have likewise observed that communicated Electricity retains the same Properties: For if a Ball of Ivory, or Wood, be set on a Glass Stand, and this Ball be render'd electrick by the Tube, it will repel all such Substances as the Tube repels; but if it be rendered electrick by applying a Cylinder of Gum-Lac near it, it will produce quite contrary Effects, viz. precisely the same as Gum-Lac would produce. In order to succeed in these Experiments, 'tis requisite that the two Bodies, which are put near one another, to find out the Nature, of their Electricity, be rendered as electrical as possible; for if one of them was not at all, or but weakly electrical, it would be attracted by the other, though it be of that Sort, that should naturally be repelled by it. But the experiment will always succeed perfectly well, if both the Bodies are sufficiently electrical. I have several other Methods to discover the Nature of the Electricity any Body is of; but my Letter is already long enough, and my Design was only to give your Grace a very succinct Extract of the Experiments I have made this Year. I beseech your Grace to communicate it to the Royal Society, and in particular to Mr. Gray, who works on this Subject with so much Application and Success, and to whom I acknowledge myself indebted for the Discoveries I have made, as well as for those I may possibly make hereafter; since 'tis from his Writings that I took the Resolution of applying my self to this kind of Experiments.

I have the Honour to be with the most sincere, and most respectuous Attachment,

My LORD,

Your GRACE:

Most Humble and most

Obedient Servant,

Du Fay.

FINIS.

Reading 15

Franklin: From Benjamin Franklin to Peter Collinson, 25 May 1747; Franklin: From Benjamin Franklin to Peter Collinson, 28 July 1747

From Benjamin Franklin to Peter Collinson, 25 May 1747

We had for some Time been of Opinion, that the Electrical Fire was not created by Friction, but collected, being an Element diffused among, and attracted by other Matter, particularly by Water and Metals. We had even discovered and demonstrated its Afflux to the Electrical Sphere, as well as its Efflux, by Means of little light Wind-Mill Wheels made of stiff Paper Vanes, fixt obliquely, and turning freely on fine Wire Axes. Also by little Wheels of the same Matter, but formed like Water Wheels. Of the Disposition and Application of which Wheels, and the various Phaenomena resulting, I could, if I had Time, and it were necessary, fill you a Sheet.

The Impossibility of Electrising one's self (tho' standing on Wax) by Rubbing the Tube and drawing the Fire from it: and the Manner of doing it by passing the Tube near a Person, or Thing standing on the Floor &c. had also occurred to us some Months before Mr. Watsons ingenious *Sequel* came to hand; and these were some of the new Things I intended to have communicated to you: But now I need only mention some Particulars not hinted in that Piece, with our Reasonings thereon; tho' perhaps the latter might well enough be spared.

1. A Person standing on Wax and rubbing the Tube; and another Person on Wax drawing the Fire, they will both of them (provided they do not stand so as to touch one another) appear to be electrised to a Person standing on the Floor; that is, he will perceive a Spark on approaching each of them.

2. But if the Persons standing on Wax touch one another during the exciting of the Tube, neither of them will appear to be electrised.

3. If they touch one another after exciting the Tube, and drawing the Fire as aforesaid, there will be a stronger Spark between them than was between either of them and the Person on the Floor.

4. After such strong Spark, neither of them discovers any Electricity.

These Appearances we attempt to account for thus. We suppose as aforesaid, That Electrical Fire is a common Element, of which every one of the three Persons abovementioned has his equal Share before any Operation is begun with the Tube. A who stands on Wax, and rubs the Tube, collects the Electrical Fire from himself into the Glass; and his Communication with the common Stock being cut off by the Wax, his Body is not again immediately supply'd. B, who stands upon Wax likewise, passing his Knuckle along near the Tube, receives the Fire which was collected by the Glass from A; and his Communication with the common Stock being likewise cutt off, he retains the additional Quantity received. To C, standing on the Floor, both appear to be electrised; for he having only the middle Quantity of Electrical Fire receives a Spark on approaching B, who has an over-quantity, but gives one to A, who has an under-quantity. If A and B touch each other, the Spark between them is stronger, because the Difference between them is greater. After such Touch, there is no Spark between either of them and C; because the Electrical Fire in all is reduced to the original Equality. If they touch while Electrising, the Equality is never destroyed, the Fire only circulating. Hence have arisen some new Terms among us. We say B (and other Bodies alike circumstanced) are electrised *positively*; A *negatively:* Or rather B is electrised *plus* and A

minus. And we daily in our Experiments electrise Bodies *plus* or *minus* as wethink proper. *These Terms* we may use till your Philosophers give us better. To electrise *plus* or*minus*, no more needs to be known than this; that the Parts of the Tube or Sphere, that are rub'd, do, in the Instant of the Friction, attract the Electrical Fire, and therefore take it from the Thing rubbing: the same Parts immediately, as the Friction upon them ceases, are disposed to give the Fire they have received, to any Body that has less. Thus you may circulate it, as Mr. Watson has shewn; You may also accumulate or subtract it upon, or from any Body, as you connect it₉ with the Rubber or with the Receiver; the Communication with the common Stock being cut off.

From Benjamin Franklin to Peter Collinson, 28 July 1747

2 At the same Time that the Wire and Top of the Bottle &c. is electrised *positively* or *plus*, the Bottom of the Bottle is electrised *negatively* or *minus* in exact Proportion. i.e. Whatever Quantity of Electrical Fire is thrown in at [the] Top, an equal Quantity goes out of the Bottom. To understand this, Suppose the common Quantity of Electricity in each Part of the Bottle, before the Operation begins, is equal to 20, and at every Stroke of the Tube, suppose a Quantity equal to 1 is thrown in; then after the first Stroke, the Quantity contained in the Wire and upper Part of the Bottle will be 21, in the Bottom 19. After the second, the upper Part will have 22, the lower 18, and so on, till after 20 Strokes, the upper Part will have a Quantity of Electrical Fire equal to 40, the lower none: and then the Operation ends; for no more can be thrown into the upper Part, when no more can be driven out of the lower Part. If you attempt to throw more in, it is spued back thro' the Wire, or flies out in loud Cracks *thro' the Sides of the Bottle*.

3 The Equilibrium can not be restored in the Bottle by *inward* Communication, or Contact of the Parts; but it must be done by a Communication form'd *without* the Bottle, between the Top and Bottom, by some Non-electric touching both at the same Time. In which Case, if the Contact be large especially, it is restored with a Violence and Quickness inexpressible; or touching each alternately, in which Case the Equilibrium is restored by Degrees.

4 As no more Electrical Fire can be thrown into the Top of the Bottle, when all is driven out of the Bottom. So in a Bottle not yet electrised, none can be thrown into the Top, when none *can* get out of the Bottom; which happens either when the Bottom is too thick, or when the Bottle is placed on an Electric-per-se. Again, when the Bottle is electrised, but little of the Electrical Fire can be *drawn out* from the Top, by touching the Wire, unless an equal Quantity can at the same Time *get in* at the Bottom. Thus place an Electris'd Bottle on clean Glass or dry Wax, and you will not, by touching the Wire, get out the Fire from the Top. Place it on a Non-electric, and touch the Wire, you will get it out in a short Time; but soonest, when you form a direct Communication, as above.

Reading 16
Faraday: Experimental Researches in Electricity

First Series.

§ 1. *On the Induction of Electric Currents.* § 2. *On the Evolution of Electricity from Magnetism.* § 3. *On a new Electrical Condition of Matter.* § 4. *On* Arago's *Magnetic Phenomena.*

[Read November 24, 1831.]

1. The power which electricity of tension possesses of causing an opposite electrical state in its vicinity has been expressed by the general term Induction; which, as it has been received into scientific language, may also, with propriety, be used in the same general sense to express the power which electrical currents may possess of inducing any particular state upon matter in their immediate neighbourhood, otherwise indifferent. It is with this meaning that I purpose using it in the present paper.

2. Certain effects of the induction of electrical currents have already been recognised and described: as those of magnetization; Ampère's experiments of bringing a copper disc near to a flat spiral; his repetition with electro-magnets of Arago's extraordinary experiments, and perhaps a few others. Still it appeared unlikely that these could be all the effects which induction by currents could produce; especially as, upon dispensing with iron, almost the whole of them disappear, whilst yet an infinity of bodies, exhibiting definite phenomena of induction with electricity of tension, still remain to be acted upon by the induction of electricity in motion.

3. Further: Whether Ampère's beautiful theory were adopted, or any other, or whatever reservation were mentally made, still it appeared very extraordinary, that as every electric current was accompanied by a corresponding intensity of magnetic action at right angles to the current, good conductors of electricity, when placed within the sphere of this action, should not have any current induced through them, or some sensible effect produced equivalent in force to such a current.

4. These considerations, with their consequence, the hope of obtaining electricity from ordinary magnetism, have stimulated me at various times to investigate experimentally the inductive effect of electric currents. I lately arrived at positive results; and not only had my hopes fulfilled, but obtained a key which appeared to me to open out a full explanation of Arago's magnetic phenomena, and also to discover a new state, which may probably have great influence in some of the most important effects of electric currents.

5. These results I purpose describing, not as they were obtained, but in such a manner as to give the most concise view of the whole.

§ 1. *Induction of Electric Currents.*

6. About twenty-six feet of copper wire one twentieth of an inch in diameter were wound round a cylinder of wood as a helix, the different spires of which were prevented from touching by a thin interposed twine. This helix was covered with calico, and then a second wire applied in the same manner. In this way twelve helices were superposed, each containing an average length of wire of twenty-seven feet, and all in the same direction. The first, third, fifth, seventh, ninth, and eleventh of these helices were connected at their extremities end to end, so as to form one helix; the others were connected in a similar manner; and thus two principal helices were produced, closely interposed, having the same direction, not touching anywhere, and each containing one hundred and fifty-five feet in length of wire.

7. One of these helices was connected with a galvanometer, the other with a voltaic battery of ten pairs of plates four inches square, with double coppers and well charged; yet not the slightest sensible reflection of the galvanometer-needle could be observed.

8. A similar compound helix, consisting of six lengths of copper and six of soft iron wire, was constructed. The resulting iron helix contained two hundred and fourteen feet of wire, the resulting copper helix two hundred and eight feet; but whether the current from the trough was passed through the copper or the iron helix, no effect upon the other could be perceived at the galvanometer.

9. In these and many similar experiments no difference in action of any kind appeared between iron and other metals.

10. Two hundred and three feet of copper wire in one length were coiled round a large block of wood; other two hundred and three feet of similar wire were interposed as a spiral between the turns of the first coil, and metallic contact everywhere prevented by twine. One of these helices was connected with a galvanometer, and the other with a battery of one hundred pairs of plates four inches square, with double coppers, and well charged. When the contact was made, there was a sudden and very slight effect at the galvanometer, and there was also a similar slight effect when the contact with the battery was broken. But whilst the voltaic current was continuing to pass through the one helix, no galvanometrical appearances nor any effect like induction upon the other helix could be perceived, although the active power of the battery was proved to be great, by its heating the whole of its own helix, and by the brilliancy of the discharge when made through charcoal.

11. Repetition of the experiments with a battery of one hundred and twenty pairs of plates produced no other effects; but it was ascertained, both at this and the former time, that the slight deflection of the needle occurring at the moment of completing the connexion, was always in one direction, and that the equally slight deflection produced when the contact was broken, was in the other direction; and also, that these effects occurred when the first helices were used (6. 8.).

12. The results which I had by this time obtained with magnets led me to believe that the battery current through one wire, did, in reality, induce a similar current through the other wire, but that it continued for an instant only, and partook more of the nature of the electrical wave passed through from the shock of a common Leyden jar than of the current from a voltaic battery, and therefore might magnetise a steel needle, although it scarcely affected the galvanometer.

13. This expectation was confirmed; for on substituting a small hollow helix, formed round a glass tube, for the galvanometer, introducing a steel needle, making contact as before between the battery and the inducing wire (7. 10.), and then removing the needle before the battery contact was broken, it was found magnetised.

14. When the battery contact was first made, then an unmagnetised needle introduced into the small indicating helix (13.), and lastly the battery contact broken, the needle was found magnetised to an equal degree apparently as before; but the poles were of the contrary kind.

15. The same effects took place on using the large compound helices first described (6. 8.).

16. When the unmagnetised needle was put into the indicating helix, before contact of the inducing wire with the battery, and remained there until the contact was broken, it exhibited little or no magnetism; the first effect having been nearly neutralised by the second (13. 14.). The force of the induced current upon making contact was found always to exceed that of the induced current at breaking of contact; and if therefore the contact was made and broken many times in succession, whilst the needle remained in the indicating helix, it at last came out not unmagnetised, but a needle magnetised as if the induced current upon making contact had acted alone on it. This effect may be due to the accumulation (as it is called) at the poles of the unconnected pile, rendering the current upon first making contact more powerful than what it is afterwards, at the moment of breaking contact.

17. If the circuit between the helix or wire under induction and the galvanometer or indicating spiral was not rendered complete *before* the connexion between the battery and the inducing wire was completed or broken, then no effects were perceived at the galvanometer. Thus, if the battery communications were first made, and then the wire under induction connected with the indicating helix, no magnetising power was there exhibited. But still retaining the latter communications, when those with the battery were broken, a magnet was formed in the helix, but of the second kind (14.), i.e. with poles indicating a current in the same direction to that belonging to the battery current, or to that always induced by that current at its cessation.

18. In the preceding experiments the wires were placed near to each other, and the contact of the inducing one with the buttery made when the inductive effect was required; but as the particular action might be supposed to be exerted only at the moments of making and breaking contact, the induction was produced in another way. Several feet of copper wire were stretched in wide zigzag forms, representing the letter W, on one surface of a broad board; a second wire was stretched in precisely similar forms on a second board, so that when brought near the first, the wires should everywhere touch, except that a sheet of thick paper was interposed. One of these wires was connected with the galvanometer, and the other with a voltaic battery. The first wire was then moved towards the second, and as it approached, the needle was deflected. Being then removed, the needle was deflected in the opposite direction. By first making the wires approach and then recede, simultaneously with the vibrations of the needle, the latter soon became very extensive; but when the wires ceased to move from or towards each other, the galvanometer-needle soon came to its usual position.

19. As the wires approximated, the induced current was in the *contrary* direction to the inducing current. As the wires receded, the induced current was in the *same* direction as the inducing current. When the wires remained stationary, there was no induced current (54.).

20. When a small voltaic arrangement was introduced into the circuit between the galvanometer (10.) and its helix or wire, so as to cause a permanent deflection of 30° or 40°, and then the battery of one hundred pairs of plates connected with the inducing wire, there was an instantaneous action as before (11.); but the galvanometer-needle immediately resumed and retained its place unaltered, notwithstanding the continued contact of the inducing wire with the trough: such was the case in whichever way the contacts were made (33.).

21. Hence it would appear that collateral currents, either in the same or in opposite directions, exert no permanent inducing power on each other, affecting their quantity or tension.

22. I could obtain no evidence by the tongue, by spark, or by heating fine wire or charcoal, of the electricity passing through the wire under induction; neither could I obtain any chemical effects, though the contacts with metallic and other solutions were made and broken alternately with those of the battery, so that the second effect of induction should not oppose or neutralise the first (13. 16.).

23. This deficiency of effect is not because the induced current of electricity cannot pass fluids, but probably because of its brief duration and feeble intensity; for on introducing two large copper plates into the circuit on the induced side (20.), the plates being immersed in brine, but prevented from touching each other by an interposed cloth, the effect at the indicating galvanometer, or helix, occurred as before. The induced electricity could also pass through a voltaic trough (20.). When, however, the quantity of interposed fluid was reduced to a drop, the galvanometer gave no indication.

24. Attempts to obtain similar effects by the use of wires conveying ordinary electricity were doubtful in the results. A compound helix similar to that already described, containing eight elementary helices (6.), was used. Four of the helices had their similar ends bound together by wire, and the two general terminations thus produced connected with the small magnetising helix containing an unmagnetised needle (13.). The other four helices were similarly arranged, but their ends connected with a Leyden jar. On passing the discharge, the needle was found to be a magnet; but it appeared probable that a part of the electricity of the jar had passed off to the small helix, and so magnetised the needle. There was indeed no reason to expect that the electricity of a jar possessing as it does great tension, would not diffuse itself through all the metallic matter interposed between the coatings.

25. Still it does not follow that the discharge of ordinary electricity through a wire does not produce analogous phenomena to those arising from voltaic electricity; but as it appears impossible to separate the effects produced at the moment when the discharge begins to pass, from the equal and contrary effects produced when it ceases to pass (16.), inasmuch as with ordinary electricity these periods are simultaneous, so there can be scarcely any hope that in this form of the experiment they can be perceived.

26. Hence it is evident that currents of voltaic electricity present phenomena of induction somewhat analogous to those produced by electricity of tension, although, as will be seen hereafter, many differences exist between them. The result is the production of other currents, (but which are only momentary,) parallel, or tending to parallelism, with the inducing current. By reference to the poles of the needle formed in the indicating helix (13. 14.) and to the deflections of the galvanometer-needle (11.), it was found in all cases that the induced current, produced by the first action of the inducing current, was in the contrary direction to the latter, but that the current produced by the cessation of the inducing current was in the same direction (19.). For the purpose of avoiding periphrasis, I propose to call this action of the current from the voltaic battery, *volta-electric induction*. The properties of the second wire, after induction has developed the first current, and whilst the electricity from the battery continues to flow through its inducing neighbour (10. 18.), constitute a peculiar electric condition, the consideration of which will be resumed hereafter (60.). All these results have been obtained with a voltaic apparatus consisting of a single pair of plates.

§ 2. *Evolution of Electricity from Magnetism.*

27. A welded ring was made of soft round bar-iron, the metal being seven-eighths of an inch in thickness, and the ring six inches in external diameter. Three helices were put round one part of this ring, each containing about twenty-four feet of copper wire one twentieth of an inch thick; they were insulated from the iron and each other, and superposed in the manner before described (6.), occupying about nine inches in length upon the ring. They could be used separately or conjointly; the group may be distinguished by the letter A (Pl. I. fig. 1.). On the other part of the ring about sixty feet of similar copper wire in two pieces were applied in the same manner, forming a helix B, which had the same common direction with the helices of A, but being separated from it at each extremity by about half an inch of the uncovered iron.

28. The helix B was connected by copper wires with a galvanometer three feet from the ring. The helices of A were connected end to end so as to form one common helix, the extremities of which were connected with a battery of ten pairs of plates four inches square. The galvanometer was immediately affected, and to a degree far beyond what has been described when with a battery of tenfold power helices *without iron* were used (10.); but though the contact was continued, the effect was not permanent, for the needle soon came to rest in its natural position, as if quite indifferent to the attached electro-magnetic arrangement. Upon breaking the contact with the battery, the needle was again powerfully deflected, but in the contrary direction to that induced in the first instance.

29. Upon arranging the apparatus so that B should be out of use, the galvanometer be connected with one of the three wires of A (27.), and the other two made into a helix through which the current from the trough (28.) was passed, similar but rather more powerful effects were produced.

30. When the battery contact was made in one direction, the galvanometer-needle was deflected on the one side; if made in the other direction, the deflection was on the other side. The deflection on breaking the battery contact was always the reverse of that produced by completing it. The deflection on making a battery contact always indicated an induced current in the opposite direction to that from the battery; but on breaking the contact the deflection indicated an induced current in the same direction as that of the battery. No making or breaking of the contact at B side, or in any part of the galvanometer circuit, produced any effect at the galvanometer. No continuance of the battery current caused any deflection of the galvanometer-needle. As the above results are common

to all these experiments, and to similar ones with ordinary magnets to be hereafter detailed, they need not be again particularly described.

31. Upon using the power of one hundred pairs of plates (10.) with this ring, the impulse at the galvanometer, when contact was completed or broken, was so great as to make the needle spin round rapidly four or five times, before the air and terrestrial magnetism could reduce its motion to mere oscillation.

32. By using charcoal at the ends of the B helix, a minute *spark* could be perceived when the contact of the battery with A was completed. This spark could not be due to any diversion of a part of the current of the battery through the iron to the helix B; for when the battery contact was continued, the galvanometer still resumed its perfectly indifferent state (28.). The spark was rarely seen on breaking contact. A small platina wire could not be ignited by this induced current; but there seems every reason to believe that the effect would be obtained by using a stronger original current or a more powerful arrangement of helices.

33. A feeble voltaic current was sent through the helix B and the galvanometer, so as to deflect the needle of the latter 30° or 40°, and then the battery of one hundred pairs of plates connected with A; but after the first effect was over, the galvanometer-needle resumed exactly the position due to the feeble current transmitted by its own wire. This took place in whichever way the battery contacts were made, and shows that here again (20.) no permanent influence of the currents upon each other, as to their quantity and tension, exists.

34. Another arrangement was then employed connecting the former experiments on volta-electric induction (6-26.) with the present. A combination of helices like that already described (6.) was constructed upon a hollow cylinder of pasteboard: there were eight lengths of copper wire, containing altogether 220 feet; four of these helices were connected end to end, and then with the galvanometer (7.); the other intervening four were also connected end to end, and the battery of one hundred pairs discharged through them. In this form the effect on the galvanometer was hardly sensible (11.), though magnets could be made by the induced current (13.). But when a soft iron cylinder seven eighths of an inch thick, and twelve inches long, was introduced into the pasteboard tube, surrounded by the helices, then the induced current affected the galvanometer powerfully and with all the phenomena just described (30.). It possessed also the power of making magnets with more energy, apparently, than when no iron cylinder was present.

35. When the iron cylinder was replaced by an equal cylinder of copper, no effect beyond that of the helices alone was produced. The iron cylinder arrangement was not so powerful as the ring arrangement already described (27.).

36. Similar effects were then produced by *ordinary magnets*: thus the hollow helix just described (34.) had all its elementary helices connected with the galvanometer by two copper wires, each five feet in length; the soft iron cylinder was introduced into its axis; a couple of bar magnets, each twenty-four inches long, were arranged with their opposite poles at one end in contact, so as to resemble a horse-shoe magnet, and then contact made between the other poles and the ends of the iron cylinder, so as to convert it for the time into a magnet (fig. 2.): by breaking the magnetic contacts, or reversing them, the magnetism of the iron cylinder could be destroyed or reversed at pleasure.

37. Upon making magnetic contact, the needle was deflected; continuing the contact, the needle became indifferent, and resumed its first position; on breaking the contact, it was again deflected, but in the opposite direction to the first effect, and then it again became indifferent. When the magnetic contacts were reversed the deflections were reversed.

38. When the magnetic contact was made, the deflection was such as to indicate an induced current of electricity in the opposite direction to that fitted to form a magnet, having the same polarity as that really produced by contact with the bar magnets. Thus when the marked and unmarked poles were placed as in fig. 3, the current in the helix was in the direction represented, P being supposed to be the end of the wire going to the positive pole of the battery, or that end towards which the zinc plates face, and N the negative wire. Such a current would have converted the cylinder into a magnet of the opposite kind to that formed by contact with the poles A and B; and such a current moves in the opposite direction to the currents which in M. Ampère's beautiful theory are considered as constituting a magnet in the position figured[1].

39. But as it might be supposed that in all the preceding experiments of this section, it was by some peculiar effect taking place during the formation of the magnet, and not by its mere virtual approximation, that the momentary induced current was excited, the following experiment was made. All the similar ends of the compound hollow helix (34.) were bound together by copper wire, forming two general terminations, and these were connected with the galvanometer. The soft iron cylinder (34.) was removed, and a cylindrical magnet, three quarters of an inch in diameter and eight inches and a half in length, used instead. One end of this magnet was introduced into the axis of the helix (fig. 4.), and then, the galvanometer-needle being stationary, the magnet was suddenly thrust in; immediately the needle was deflected in the same direction as if the magnet had been formed by either of the two preceding processes (34. 36.). Being left in, the needle resumed its first position, and then the magnet being withdrawn the needle was deflected in the opposite direction. These effects were not great; but by introducing and withdrawing the magnet, so that the impulse each time should be added to those previously communicated to the needle, the latter could be made to vibrate through an arc of 180° or more.

40. In this experiment the magnet must not be passed entirely through the helix, for then a second action occurs. When the magnet is introduced, the needle at the galvanometer is deflected in a certain direction; but being in, whether it be pushed quite through or withdrawn, the needle is deflected in a direction the reverse of that previously produced. When the magnet is passed in and through at one continuous motion, the needle moves one way, is then suddenly stopped, and finally moves the other way.

Fourteenth Series.

§ 20. *Nature of the electric force or forces.* § 21. *Relation of the electric and magnetic forces.* § 22. *Note on electrical excitation.*

Received June 21, 1838.—Read June 21, 1838.

§ 20. *Nature of the electric force or forces.*

1667. The theory of induction set forth and illustrated in the three preceding series of experimental researches does not assume anything new as to the nature of the electric force or forces, but only as to their distribution. The effects may depend upon the association of one electric fluid with the particles of matter, as in the theory of Franklin, Epinus, Cavendish, and Mossotti; or they may depend upon the association of two electric fluids, as in the theory of Dufay and Poisson; or they may not depend upon anything which can properly be called the electric fluid, but on vibrations or other affections of the matter in which they appear. The theory is unaffected by such differences in the mode of viewing the nature of the forces; and though it professes to perform the important office of stating *how* the powers are arranged (at least in inductive phenomena), it does not, as far as I can yet perceive, supply a single experiment which can be considered as a distinguishing test of the truth of any one of these various views,

1668. But, to ascertain how the forces are arranged, to trace them in their various relations to the particles of matter, to determine their general laws, and also the specific differences which occur under these laws, is as important as, if not more so than, to know whether the forces reside in a fluid or not; and with the hope of assisting in this research, I shall offer some further developments, theoretical and experimental, of the conditions under which I suppose the particles of matter are placed when exhibiting inductive phenomena.

1669. The theory assumes that all the *particles*, whether of insulating or conducting matter, are as wholes conductors.

1670. That not being polar in their normal state, they can become so by the influence of neighbouring charged particles, the polar state being developed at the instant, exactly as in an insulated conducting *mass* consisting of many particles.

1671. That the particles when polarized are in a forced state, and tend to return to their normal or natural condition.

1672. That being as wholes conductors, they can readily be charged, either *bodily* or *polarly*.

1673. That particles which being contiguous[332] are also in the line of inductive action can communicate or transfer their polar forces one to another *more* or *less* readily.

1674. That those doing so less readily require the polar forces to be raised to a higher degree before this transference or communication takes place.

1675. That the *ready* communication of forces between contiguous particles constitutes *conduction*, and the *difficult* communication *insulation*; conductors and insulators being bodies whose particles naturally possess the property of communicating their respective forces easily or with difficulty; having these differences just as they have differences of any other natural property.

1676. That ordinary induction is the effect resulting from the action of matter charged with excited or free electricity upon insulating matter, tending to produce in it an equal amount of the contrary state.

4125 1677. That it can do this only by polarizing the particles contiguous to it, which perform the same office to the next, and these again to those beyond; and that thus the action is propagated from the excited body to the next conducting mass, and there renders the contrary force evident in consequence of the effect of communication which supervenes in the conducting mass upon the polarization of the particles of that body (1675.).

4130 1678. That therefore induction can only take place through or across insulators; that induction is insulation, it being the necessary consequence of the state of the particles and the mode in which the influence of electrical forces is transferred or transmitted through or across such insulating media.

1679. The particles of an insulating dielectric whilst under induction may be compared to a series of small magnetic needles, or more correctly still to a series of small insulated conductors. If the space
4135 round a charged globe were filled with a mixture of an insulating dielectric, as oil of turpentine or air, and small globular conductors, as shot, the latter being at a little distance from each other so as to be insulated, then these would in their condition and action exactly resemble what I consider to be the condition and action of the particles of the insulating dielectric itself (1337.). If the globe were charged, these little conductors would all be polar; if the globe were discharged, they would all
4140 return to their normal state, to be polarized again upon the recharging of the globe. The state developed by induction through such particles on a mass of conducting mutter at a distance would be of the contrary kind, and exactly equal in amount to the force in the inductric globe. There would be a lateral diffusion of force (1224. 1297.), because each polarized sphere would be in an active or tense relation to all those contiguous to it, just as one magnet can affect two or more magnetic
4145 needles near it, and these again a still greater number beyond them. Hence would result the production of curved lines of inductive force if the inducteous body in such a mixed dielectric were an uninsulated metallic ball (1219. &c.) or other properly shaped mass. Such curved lines are the consequences of the two electric forces arranged as I have assumed them to be: and, that the inductive force can be directed in such curved lines is the strongest proof of the presence of the two
4150 powers and the polar condition of the dielectric particles.

1680. I think it is evident, that in the case stated, action at a distance can only result through an action of the contiguous conducting particles. There is no reason why the inductive body should polarize or affect *distant* conductors and leave those *near* it, namely the particles of the dielectric, unaffected: and everything in the form of fact and experiment with conducting masses or particles
4155 of a sensible size contradicts such a supposition.

1681. A striking character of the electric power is that it is limited and exclusive, and that the two forces being always present are exactly equal in amount. The forces are related in one of two ways, either as in the natural normal condition of an uncharged insulated conductor; or as in the charged state, the latter being a case of induction.

4160 1682. Cases of induction are easily arranged so that the two forces being limited in their direction shall present no phenomena or indications external to the apparatus employed, Thus, if a Leyden jar, having its external coating a little higher than the internal, be charged and then its charging ball and rod removed, such jar will present no electrical appearances so long as its outside is uninsulated. The two forces which may be said to be in the coatings, or in the particles of the dielectric contiguous to
4165 them, are entirely engaged to each other by induction through the glass; and a carrier ball (1181.) applied either to the inside or outside of the jar will show no signs of electricity. But if the jar be

insulated, and the charging ball and rod, in an uncharged state and suspended by an insulating thread of white silk, be restored to their place, then the part projecting above the jar will give electrical indications and charge the carrier, and at the same time the *outside* coating of the jar will be found in the opposite state and inductric towards external surrounding objects.

1683. These are simple consequences of the theory. Whilst the charge of the inner coating could induce only through the glass towards the outer coating, and the latter contained no more of the contrary force than was equivalent to it, no induction external to the jar could be perceived; but when the inner coating was extended by the rod and ball so that it could induce through the air towards external objects, then the tension of the polarized glass molecules would, by their tendency to return to the normal state, fall a little, and a portion of the charge passing to the surface of this new part of the inner conductor, would produce inductive action through the air towards distant objects, whilst at the same time a part of the force in the outer coating previously directed inwards would now be at liberty, and indeed be constrained to induct outwards through the air, producing in that outer coating what is sometimes called, though I think very improperly, free charge. If a small Leyden jar be converted into that form of apparatus usually known by the name of the electric well, it will illustrate this action very completely.

1684. The terms *free charge* and *dissimulated electricity* convey therefore erroneous notions if they are meant to imply any difference as to the mode or kind of action. The charge upon an insulated conductor in the middle of a room is in the same relation to the walls of that room as the charge upon the inner coating of a Leyden jar is to the outer coating of the same jar. The one is not more *free* or more *dissimulated* than the other; and when sometimes we make electricity appear where it was not evident before, as upon the outside of a charged jar, when, after insulating it, we touch the inner coating, it is only because we divert more or less of the inductive force from one direction into another; for not the slightest change is in such circumstances impressed upon the character or action of the force.

* * * * *

1685. Having given this general theoretical view, I will now notice particular points relating to the nature of the assumed electric polarity of the insulating dielectric particles.

1686. The polar state may be considered in common induction as a forced state, the particles tending to return to their normal condition. It may probably be raised to a very high degree by approximation of the inductric and inducteous bodies or by other circumstances; and the phenomena of electrolyzation (861. 1652. 1796.) seem to imply that the quantity of power which can thus be accumulated on a single particle is enormous. Hereafter we may be able to compare corpuscular forces, as those of gravity, cohesion, electricity, and chemical affinity, and in some way or other from their effects deduce their relative equivalents; at present we are not able to do so, but there seems no reason to doubt that their electrical, which are at the same time their chemical forces (891. 918.), will be by far the most energetic.

§ 21. *Relation of the electric and magnetic forces.*

1709. I have already ventured a few speculations respecting the probable relation of magnetism, as the transverse force of the current, to the divergent or transverse force of the lines of inductive action belonging to static electricity (1658, &c.).

1710. In the further consideration of this subject it appeared to me to be of the utmost importance to ascertain, if possible, whether this lateral action which we call magnetism, or sometimes the induction of electrical currents (26. 1048, &c.), is extended to a distance *by the action of the intermediate particles* in analogy with the induction of static electricity, or the various effects, such as conduction, discharge, &c., which are dependent on that induction; or, whether its influence at a distance is altogether independent of such intermediate particles (1662.).

* * * * *

1731. If the lateral or transverse force of electrical currents, or what appears to be the same thing, magnetic power, could be proved to be influential at a distance independently of the intervening contiguous particles, then, as it appears to me, a real distinction of a high and important kind, would be established between the natures of these two forces (1654. 1664.). I do not mean that the powers are independent of each other and might be rendered separately active, on the contrary they are probably essentially associated (1654.), but it by no means follows that they are of the same nature. In common statical induction, in conduction, and in electrolyzation, the forces at the opposite extremities of the particles which coincide with the lines of action and have commonly been distinguished by the term electric, are polar, and in the cases of contiguous particles act only to insensible distances; whilst those which are transverse to the direction of these lines, and are called magnetic, are circumferential, act at a distance, and if not through the mediation of the intervening particles, have their relations to ordinary matter entirely unlike those of the electrical forces with which they are associated.

1732. To decide this question of the identity or distinction of the two kinds of power, and establish their true relation, would be exceedingly important. The question seems fully within the reach of experiment, and offers a high reward to him who will attempt its settlement.

1733. I have already expressed a hope of finding an effect or condition which shall be to statical electricity what magnetic force is to current electricity (1658.). If I could have proved to my own satisfaction that magnetic forces extended their influence to a distance by the conjoined action of the intervening particles in a manner analogous to that of electrical forces, then I should have thought that the natural tension of the lines of inductive action (1659.), or that state so often hinted at as the electro-tonic state (1661. 1662.), was this related condition of statical electricity.

1734. It may be said that the state of *no lateral action* is to static or inductive force the equivalent of *magnetism* to current force; but that can only be upon the view that electric and magnetic action are in their nature essentially different (1664.). If they are the same power, the whole difference in the results being the consequence of the difference of *direction*, then the normal or *undeveloped* state of electric force will correspond with the state of *no lateral action* of the magnetic state of the force; the electric current will correspond with the lateral effects commonly called magnetism; but the state of static induction which is between the normal condition and the current will still require a

4245 corresponding lateral condition in the magnetic series, presenting its own peculiar phenomena; for it can hardly be supposed that the normal electric, and the inductive or polarized electric, condition, can both have the same lateral relation. If magnetism be a separate and a higher relation of the powers developed, then perhaps the argument which presses for this third condition of that force would not be so strong.

Reading 17
Einstein: On the Electrodynamics of moving bodies

Translated by Megnad Saha

It is known that the application of MAXWELL's electrodynamics, as ordinarily conceived at the present time, to moving bodies, leads to asymmetries which don't seem to be connected with the phenomena. Let us, for example, think of the mutual action between a magnet and a conductor. The observed phenomenon in this case depends only on the relative motion of the conductor and the magnet, while according to the usual conception, a strict distinction must be made between the cases where the one or the other of the bodies is in motion. If, for example, the magnet moves and the conductor is at rest, then an electric field of certain energy-value is produced in the neighbourhood of the magnet, which excites a current in those parts of the field where a conductor exists. But if the magnet be at rest and the conductor be set in motion, no electric field is produced in the neighbourhood of the magnet, but an electromotive force is produced in the conductor which corresponds to no energy *per se*; however, this causes — equality of the relative motion in both considered cases is assumed — an electric current of the same magnitude and the same course, as the electric force in the first case.

Examples of a similar kind, as well as the unsuccessful attempts to substantiate the motion of the earth relative to the "light-medium", lead us to the supposition that not only in mechanics, but also in electrodynamics, no properties of the phenomena correspond to the concept of absolute rest, but rather that for all coordinate systems for which the mechanical equations hold, the equivalent electrodynamical and optical equations hold also, as has already been shown for magnitudes of the first order. In the following we will elevate this guess to a presupposition (whose content we shall subsequently call the "Principle of Relativity") and introduce the further assumption, — an assumption which is only apparently irreconcilable with the former one — that light in empty space always propagates with a velocity V which is independent of the state of motion of the emitting body. These two assumptions are quite sufficient to give us a simple and consistent theory of electrodynamics of moving bodies on the basis of the **Maxwellia**n theory for bodies at rest. The introduction of a **"luminiferous æther"** will be proved to be superfluous in so far, as according to the conceptions which will be developed, we shall introduce neither a "space absolutely at rest" endowed with special properties, nor shall we associate a velocity-vector with a point in which electro-magnetic processes take place.

Like every other theory in electrodynamics, the theory to be developed is based on the kinematics of rigid bodies; since in the arguments of every theory, we have to do with relations between rigid bodies (co-ordinate system), clocks, and electromagnetic processes. An insufficient consideration of these circumstances is the cause of difficulties with which the electrodynamics of moving bodies has to fight at present.

I. Kinematical Part

§ 1. Definition of Simultaneity.

Let us have a co-ordinate system, in which the NEWTONian equations hold. For verbally distinguishing this system from another which will be introduced hereafter, and for clarification of the idea, we shall call it the "stationary system."

If a material point be at rest in this coordinate-system, then its position in this system can be found out by a measuring rod by using the methods of Euclidean Geometry, and can be expressed in Cartesian co-ordinates.

If we wish to describe the *motion* of a material point, the values of its coordinates must be expressed as functions of time. Now, it is always to be borne in mind that such a mathematical definition has a physical meaning only, when we have a clear notion of what is meant by "time". We have to take into consideration the fact that those of our assessments, in which time plays a role, are always judgements on *simultaneous events*. For example, we say "that a train arrives here at 7 o'clock"; this means "that the exact pointing of the little hand of my watch to 7, and the arrival of the train are simultaneous events".

It may appear that all difficulties connected with the definition of "time" can be removed when in place of "time", we substitute "the position of the little hand of my watch". Such a definition is in fact sufficient, when it is required to define time exclusively for the place at which the clock is stationed. But the definition is not sufficient when it is required to chronologically connect a series of events taking place at different places — or what amounts to the same thing — to chronologically evaluate the occurrence of events, which take place at places distant from the clock.

However, to chronologically evaluate the events, we can satisfy ourselves by assuming an observer who is stationed at the origin of coordinates with the clock, and who associates a signal of light – giving testimony of the event to be estimated and which comes to him through empty space – with the corresponding position of the hands of the clock. Deficiency of such an association is — as we know by experience — that it depends on the position of the observer provided with the clock. We can attain a far more practical result by the following treatment.

If an observer be stationed at A with a clock, he can estimate the time of events occurring in the immediate neighbourhood of A by looking for the position of the hands of the clock, which are simultaneous with the event. If a clock be stationed also at point B in space, — we should add that "the clock is exactly of the same nature as the one at A", — then the chronological evaluations of the events occurring in the immediate vicinity of B, is possible for an observer located in B. But without further premises, it is not possible to chronologically compare the events at B with the events at A. We have hitherto only an "A-time", and a "B-time", but no "time" common to A and B. This last time (*i.e.*, common time) can now be defined, however, if we establish *by definition* that the "time" which light requires in travelling from A to B is equivalent to the "time" which light requires in travelling from B to A. For example, a ray of light proceeds from A at "A-time" t_A towards B, arrives and is reflected from B at B-time t_B, and returns to A at "A-time" t'_A. According to the definition, both clocks are synchronous, if

$$t_B - t_A = t'_A - t_B$$

We assume that this definition of synchronism is possible without involving any inconsistency, for any number of points, therefore the following relations hold:

1. If the clock at B be synchronous with the clock at A, then the clock at A is synchronous with the clock at B.

2. If the clock at A be synchronous with the clock at B as well as with the clock at C, then also the clocks at B and C are synchronous.

Thus with the help of certain (imagined) physical experiences, we have established what we understand when we speak of clocks at rest at different places, and synchronous with one another; and thereby we have arrived at a definition of "synchronism" and "time". The "time" of an event is the simultaneous indication of a stationary clock located at the place of the event, which is synchronous with a certain stationary clock for all time determinations.

In accordance with experience we shall assume that the magnitude

$$\frac{2\overline{AB}}{t'_A - t_A} = V,$$

is a universal constant (the speed of light in empty space).

We have defined time essentially with a clock at rest in a stationary system. On account of its affiliation to the stationary system, we call the time defined in this way "the time of the stationary system."

§ 2. On the relativity of lengths and times.

The following considerations are based on the Principle of Relativity and on the Principle of Constancy of the velocity of light, both of which we define in the following way.

1. The laws according to which the states of physical systems alter are independent of the choice, to which of two co-ordinate systems (having a uniform translatory motion relative to each other) these state changes are related.

2. Every ray of light moves in the "stationary" co-ordinate system with the definite velocity V, the velocity being independent of the condition, whether this ray of light is emitted by a body at rest or in motion. Here

$$\text{velocity} = \frac{\text{Path of Light}}{\text{Interval of time}}$$

where "interval of time," is to be understood as defined in § 1.

Let us have a rigid stationary rod; it has a length l, when measured by a measuring rod also at rest. We shall assume that the axis of the rod is the X-axis of the stationary coordinate-system. Let us now impart to the rod a uniform velocity v, parallel to the axis of X and in the increasing direction of x. What is the length of the *moving* rod; this can be thought as obtained by either of these operations:

a) The observer provided with the measuring rod moves along with the rod to be measured, and measures by direct superposition the length of the rod, just as if the observer, the measuring rod, and the rod to be measured were at rest.

b) The observer measures, at which points of the stationary system the ends of the rod to be measured are located at a particular time t, by means of clocks placed in the stationary system (the clocks being synchronous as defined in § 1). The distance between these two points, measured by the previously used measuring rod, this time it being at rest, is also a length, which we may call the "length of the rod."

According to the Principle of Relativity, the length found out by the operation a), which we may call "the length of the rod in the moving system", is equal to the length l of the stationary rod.

The length which is found out by operation b), may be called "the length of the (moving) rod in the stationary system". This length is to be estimated on the basis of our two principles, and we shall find it to be different from l.

In the generally employed kinematics, we tacitly assume that the lengths defined by these two operations are equal, or in other words, that at an epoch of time t, a moving rigid body is geometrically replaceable by *the same* body, when it is *at rest* in a particular location.

Let us furthermore suppose that the two clocks synchronous with the clocks in the system at rest are brought to the ends A and B of a rod, *i.e.*, the indications of the clocks correspond to the "time of the stationary system" at the places where they happen to arrive; these clocks are therefore "synchronous in the stationary system".

We further imagine that there are two observers at the two clocks, and moving with them, and that these observers apply the criterion for synchronism stated in §1, to the two clocks. At the time t_A, a ray of light goes out from A, is reflected from B at the time t_B, and arrives back at A at time t'_A. Taking into consideration the principle of constancy of the velocity of light, we have

$$t_B - t_A = \frac{r_{AB}}{V - v}$$

and

$$t'_A - t_B = \frac{r_{AB}}{V + v},$$

where r_{AB} is the length of the moving rod, measured in the stationary system. Therefore the observers moving with the moving rod, thus would not find the clocks synchronous, though the observers in the stationary system would declare the clocks to be synchronous.

We therefore see that we can attach no *absolute* significance to the concept of synchronism; two events which are synchronous when viewed from one coordinate-system, will not be synchronous when viewed from a system moving relatively to this system.

www.ingramcontent.com/pod-product-compliance
Lightning Source LLC
Chambersburg PA
CBHW080921170526
45158CB00008B/2188